Agrarian Environments

Agrarian Environments

Resources, Representations, and Rule in India

Edited by Arun Agrawal and K. Sivaramakrishnan

Foreword by James C. Scott

Duke University Press *Durham & London 2000*

© 2000 Duke University Press
All rights reserved
Printed in the United States of
America on acid-free paper ∞
Designed by Rebecca M. Giménez
Typeset in Carter & Cone Galliard
by Tseng Information Systems, Inc.
Library of Congress Cataloging-
in-Publication Data appear on the
last printed page of this book.

Contents

Labored Landscapes: Agro-ecological Change in Central Gujarat, India
Vinay Gidwani 216

REFLECTIONS

Agrarian Histories and Grassroots Development in South Asia
David Ludden 251

Cathecting the Natural *Ajay Skaria* 265

Bibliography 277

Contributors 303

Index 307

Foreword *James C. Scott*

Being asked to write a brief foreword to a collection of this quality and breadth would always be a privilege. The honor, in this particular case, is enhanced in at least three ways. First, I believe that the work assembled here and the conference in which the papers were discussed represent the groundwork for an important intellectual advance in our thinking about environment and agriculture. Second, K. Sivaramakrishnan and Arun Agrawal, the intellectual progenitors and editors of this volume, are themselves the authors of distinguished work in this "third wave" of environmental analysis—some of it already published, some "in the pipeline." Third, the Program in Agrarian Studies at Yale, which I helped found, can legitimately claim to have been the matchmaker for this enviable collaboration. If there is one principle for which the Program in Agrarian Studies stands, it is the pathbreaking, grounded, interdisciplinary work found between these covers.

As I read these papers I came to think of them as the third generation of environmental discourse on South Asia. That is no small achievement, inasmuch as the environmental and agrarian literature about the subcontinent has, as with subaltern studies, so often set the intellectual and conceptual tone for work on comparable issues elsewhere in the world. There are any number of rigid categories, binary distinctions, and ahistorical truisms that either do not survive this volume or, at best, emerge severely recast and qualified in the light of this new work.

Among the great services this volume performs is to demonstrate the artificiality of such categories as arable, forest, pasture, et cetera, as well as categories of livelihoods based on them: cultivation, hunting-gathering, pastoralism. The movement within and between such categories, unclassifiable mixed cases, the strong interdependence between various modes of livelihood, and the radical changes over time in landscape, markets, climate, and human strategies of land use defy such simple distinctions. These categories were not, of course, merely the stock-in-trade of intellectuals and ethnographers; they were also applied administrative categories that marked the entire history of India—precolonial, colonial, and independent. The contributors to *Agrarian Environments* show, again and again, the historical contingency of such boundaries of thought and practice—continually breached by adaptive practice, social uncertainties, and the play of political power.

In earlier environmental work about the subcontinent, one can discern the outlines of what I have heard called the "standard narrative." While the work of the most gifted of this "second generation" was too rich to be entirely characterized by its simplest version, it was and is nevertheless still in daily polemical use. According to the crudest formulation of this "narrative," the colonizers, the market, and the state were the agents of ecological degradation while indigenous peoples, the more neolithic the better, were nature's natural conservators. It is a narrative that has been exported or independently invented almost everywhere.

The work in this volume does not demolish this account entirely, but it refuses to take it for granted. It rejects this standard account as an a priori assumption and insists on the violence it does to a history and ecology of such bewildering variety and contingency as to escape such a simple formula. If this volume has an implicit methodological commitment, it is that only historically situated, empirically grounded accounts of the interaction between human activity and environment (an environment that is always anthropogenic), accounts that are agnostic about the received categories of landscape and behavior, can possibly advance our knowledge. The result is a series of site-specific, historically situated studies that manage always to remain theoretically and conceptually self-conscious.

Having seen the contributors to *Agrarian Environments* revise and transcend the work of an earlier generation, it remains, perhaps, for me to point out that the earlier work was the precondition—the compost, if you will—from which their new growth springs. Their advances will, in turn, become the soil for new work to come—a fourth generation?—not only in South Asia but elsewhere too, given its importance as a new point of departure.

Acknowledgments

The idea for this book came into being when Arun Agrawal was a fellow at the Program in Agrarian Studies at Yale University (1995–1996), and K. Sivaramakrishnan was finishing his Ph.D. thesis in the Department of Anthropology at Yale. Two panels of papers on related themes at the Association for Asian Studies meetings in Honolulu, Hawaii, in spring 1996, and at the South Asian Studies meetings in Wisconsin in fall 1996, acted as preliminary meetings where we crystallized the idea of this volume. We thank all the panel participants and the audience members for their instructive comments and stimulating ideas. We especially acknowledge Akhil Gupta's contribution in an early stage of the project.

We didn't know it when we first started talking about the book, but K. Sivaramakrishnan was himself about to become a program fellow in 1996–1997. Surely the program can be considered to have sired the book, and not just because both the editors have been program fellows. We also received a timely and generous conference grant from the program. It allowed most of the authors to come together for a small workshop in May 1997 to discuss the individual papers and meet their discussants. Two of the contributions from the discussants appear at the end of the book. Our special thanks go to David Ludden, John Richards, and Ajay Skaria, who not only provided rich and thought-provoking commentaries on the papers at the workshop but also crafted their critiques in such a way as to have an independent force related to the idea of agrarian environments.

During the discussions in the workshop we also realized how the reification of social identities related to human-nature relationships is itself a product of the separation of the agrarian and the environmental as independent domains. In this sense, the chief argument of this volume is without doubt a collective inspiration. All the authors in the volume must share in the praise. Nor do we absolve them of any criticisms of the ideas in it; those who are party to collective inspirations must also share the burden of well-founded criticisms. For additional comments on the introduction to the volume, we would like to thank Clark Gibson, Tania Li, Donald Moore, and Ramachandra Guha. We would also like to thank the three anonymous reviewers at Duke University Press and Oxford University Press (Delhi) for having gone through the manuscript carefully. Their comments have made this volume a far more integrated set of essays.

We would like to express our gratitude also for the consistent encouragement James Scott and Kay Mansfield have provided us from the very beginning of this project. James Scott's influence, never imposed, is an inspiring model of how to think and conduct collaborative projects. Thanks, as well, to Kay Mansfield for patiently and thoroughly copyediting the manuscript. We are further grateful to Jan Opdyke for proofreading and Ylva Hernlund for help in preparing the index. Arun Agrawal thanks the National Science Foundation (grant # SBR 9905443) and the MacArthur Foundation (grant # 96-42825-WER) for additional support while working on this project. This is also the appropriate moment to thank Valerie Millholland at Duke University Press and Bela Malik at Oxford University Press (Delhi) for their support of the project, and of the ideas in it. Without their understanding of how edited volumes run into unexpected delays, and their willingness to accommodate such delays, the project would not have seen completion.

Introduction: Agrarian Environments

As *raika* shepherds and other migrant pastoralists in western Rajasthan travel between pastures, forests, and fallow, they are moving across landscapes that have a long history of changing vegetation, human activities, and state policies (Agrawal 1999). When the gregarious *sal* trees in southwest Bengal colonize an abandoned mango grove or enter a fallow upland, we are reminded that lands in this region change from cultivated fields to naturally regenerating woodlands across space and over time (Sivaramakrishnan 1996). The expansion of irrigation in some parts of Kheda, Gujarat, and its contraction in other adjoining areas is a consequence of a combination of economic and ecological factors (Gidwani 1996). These are examples of "agrarian environments." They draw attention to the blurred boundaries between an autonomous nature that supposedly stands outside of human endeavor, and a human agency that is presumed to construct all landscapes. As changing, hybrid landscapes, these are but three of the many regions in India that remind us of the prodigious energy necessary to fix the agrarian and the environmental as separate domains of existence and analysis.

Our use of the term "agrarian environments" denotes an insistent attention to a field of social negotiations around the environment in predominantly agrarian contexts. The interactions and processes we examine are unavoidably inflected by the agrarian affiliations of the actors and issues involved. They demonstrate the pervasive links between the agrarian and the environmental and suggest that to treat one independently of the other is to fail to understand either. In the predominantly agrarian socioeconomic context of India, studies that do not explore the connections of environmental changes with agrarian structures and processes delink environmental politics from the agrarian world that is both the locus and the object of these politics.

Over the last fifty years, the science of ecology, the struggles of environmental activists, and the scholarship of environmental historians have combined to demarcate the physical and conceptual field within which their concerns are to be located. Through a focus on natural resources, especially forests and water, and the analytical presumption that these resources degrade because of human impact, the field of environmental studies has created a site for itself that is resolutely separated from the agrarian world. The establishment of environmental studies has thus been predicated on a critique of scholarship and politics that were earlier preoccupied with the urban or the arable.

But the very category "environment" comes into existence only because many scholars identify the environment with "nature,"[1] remove it from the world of agrarian relations, and imagine it, ideally, as something that exists separately from humans.[2] Here scholars are producing what Deleuze and Guattari (1988, 362–64) call state science, something that imposes "an order of reasons" on the unruly nomadic sciences, or what Foucault ([1994] 1997, 73) called biopolitics, the endeavor "to rationalize the problems presented to governmental practice." The "order of reasons," in relation to environmental and agrarian studies, has the specific spatial consequence of striation, manifest as a partitioning of landscapes into distinct domains of natural existences and productive economic relations.

The support for such separation of the natural from the human has far deeper historical roots than just the foundations of environmental studies or environmental history. Using creative interpretations of the biblical Genesis story, and anthropological and economic speculation, the attempts by Hobbes, Locke, Hume, and Rousseau to think about political order and property depended on an original assumption that nature exists separately from humans.[3] The ontological status of the "environment" has thus depended on the belief in a distant past when humans had little technology, and, hence, scant ability to transform the natural into the artificial.[4]

Over time the pristine character of nature could be imagined in varied ways, some opposed to each other. It has been envisioned as riotous and chaotic, needing the hand of man for systematic organization and productive utilization. It has also been seen as representing Edenic bliss, before man's actions led to a Fall from which it is impossible to recover. But in each of these visions, humans and the environment are distinct entities who act on each other rather than being mutually constituted.

The space in which the environment came to be constructed as the natural was created by a prior history of treating agrarian landscapes as the product of culture. The classic themes of agrarian studies—migration, commercialization, tenurial relations, changing patterns and intensities of cropped staples, credit, state formation, market and trade relations—were all directed at describing the complex social construction of nature through which agrarian societies emerged, transformed, and declined. This point can be amplified by looking at certain key areas of investigation that existing scholarship could have seized to apprehend agrarian environments as we discuss them.

AGRARIAN ENVIRONMENTAL HISTORIES AND POLITICS

Historical work on the period between the decline of the Mughal empire and the consolidation of British rule is one such example. It is ironic that agrarian historians, arguing against older assumptions of eighteenth-century chaos, have reinforced the focus on continued agricultural commodity production even in a period of major political upheaval and transitions.[5] In so doing, they missed the opportunity to consider the agrarian and the environmental together. Another example is the years between the world wars. The work of George Blyn on the agricultural statistics of late colonial India was enabled by the focus on crop production that had informed their collection, and promoted anxieties about declining food crop performance as populations rose after 1921.[6] These concerns with agricultural production powerfully framed postindependence debates on the agrarian economy. By the 1960s, the selective implementation of green revolution programs in well-endowed regions worked to heighten the divide between arable and other lands. State science, and scholarship that conformed to it, etched deeper the divisions between agrarian and environmental domains in rural India.[7]

The divided focus on landscapes partitioned as cultural and natural has ensured abiding differences in the research agendas of agrarian and environmental studies in India. Agrarian studies has concentrated primarily on regions where agricultural productivity was high, and where greater intensification of agriculture was evident. Thus the fertile Indo-Gangetic plains have received the greatest analytical notice. Even in the Deccan and the south, irrigated agriculture in river valleys and coastal plains has attracted the most attention from historical and contemporary agrarian scholarship. The presumption guiding the scholarly focus, it can be argued, has been that these regions are most capable of producing a social surplus, and that these agrarian spaces have undergone the most significant social and political changes.

Studies of agrarian change in the forty years since independence have been preoccupied intensely with the green revolution, with new technologies, with farm size, and with the political economy of state intervention in the ecologically and infrastructurally better equipped regions. Peripheral regions in the mountains, in the western semiarid plains and the hills, in the northeast, or even in the Chhotanagpur Plateau, have for the most part been ignored by scholars of agrarian politics and history. Agrarian studies has made few excursions out from the arable "heartlands" of India. Environmental scholarship, coming into its own only in the last two decades, has, conversely, been preoccupied with mountains, forests, tribal populations, and the semiarid parts in western India. Rarely has it ventured into the plains.

The regional focus of agrarian studies is perhaps derived from its theoretical interests. The study of agrarian change has been dominated by debates about transitions to capitalism, the role of commodity and credit markets, the building of agrarian empires on the revenues derived from agricultural surpluses, the impact of technological innovations such as plowing in earlier times and mechanization in more recent periods, and the social consequences of privatization. The desire to identify modes of production, modes of power, and their attendant relations of production prevented agrarian studies from looking at the environment. The attempt to pinpoint the conditions under which different modes articulated, or were superseded, drew agrarian studies to regions where these phenomena appeared most developed. The sharply polemical oppositions that emerged within south Asian studies, between elitist and subaltern scholarship in the 1980s, did not result in a critique of the separation of the agrarian

4 *Arun Agrawal and K. Sivaramakrishnan*

from the environment. In part this might be because even subaltern studies remained caught up in debates about semifeudal remnants, precapitalist community, and other such symbols of imperfect class formation in rural India.[8]

The power of typologies developed in agrarian studies reverberates in the emergent field of environmental studies. Gadgil and Guha's pioneering work (1992) illustrates that even for environmental historians, the spatial distribution of agrarian and nonagrarian social formations is adequately mapped by hunting-gathering, nomadic pastoralist, settled agricultural, and industrial modes of production. Their book begins with a discussion of modes of production and amplifies it into their concept of "modes of resource use," extending "the realm of production to include flora, fauna, water, and minerals" (Gadgil and Guha 1992, 13).

In the face of this formidable analytical legacy from agrarian as well as environmental studies, our quest to identify and explore the conceptual landscape of agrarian environments begins with some of the most recent scholarship mapping this hybrid domain in rural India. The blurred boundaries that our work consistently finds between the agrarian and the environmental suggests that typologies separating environmental and agrarian studies serve both poorly. Where farmers become pastoralists in response to regional development policies, when forest conflicts signal the workings of a regional agroforestry system, when relations between tribal groups and agriculturists are redefined in the context of changing state strategies to exercise political control, then choosing between an agrarian or an environmental perspective is not just unsatisfying but plainly misleading. It is misleading because it is a truncation and a misrepresentation of the interwoven dynamics between the agrarian and the environmental worlds. It is also misleading because it prevents an examination of ideal-typical constructs that have formed the building blocks of much agrarian and environmental research.

Consider overgrazing as an example. It is seen by many to be an ecological problem created by shortsighted pastoralists intent on increasing herds to unsustainable size. Framed thus, pastoralism becomes a discrete mode of production, capable of self-reproduction independent of any relationships with agriculture, farmers, or the arable. But the world over, and certainly in India, pastoralist livelihoods depend on interactions in the market, with farmers, and around agricultural production. Grazing lands have declined owing to the spread of settled agriculture and irrigation. Pastoralists have discovered new adaptive social-ecological niches, often in intimate contiguity with agriculturists. To talk about pastoralists, then, without considering their links with the agrarian world is to posit a model of pastoralism that is descriptively incomplete, analytically deceptive, and of limited practical use.

Agrarian environments, the chapters of this volume insist, have to be comprehended as being part of a biophysical and social environment that always includes the urban and the nonurban, the arable and the nonarable, and other areas that are integrally linked to the world of agriculture and environment and their allied social-economic relations. Not only must we reject the tunnel vision of agrarian studies, focused as it remains on the bounded terrain of river valleys and coastal plains, but we need also to move beyond the compartmentalized perspectives of the first generation of environmentalism. In the last thirty years, air, water, forests, pastures, fisheries, and wildlife have taken shape as distinct realms in nature, shored up by their separate, elaborate, legal-institutional structures. We must learn how to navigate across these domains in the search to learn more about our subject matter and the problems that interest us. As a recent review of environmental policy in the United States suggests, we should strive for "an *ecologicalism* that recognizes the inherent interdependence of all life-systems" (Esty and Chertow 1997, 45).

But we do not merely emphasize the need to see the systemic interconnectedness of rural life-support systems. In speaking of agrarian environments, and in explicitly referring to the constructed nature of all environments, we draw attention to how nature and landscapes mutate over time in their physical characteristics, human interactions, and cultural representations. Our emphasis on the malleability of landscapes is not simply for the sake of scoring a postmodern point. Rather, the attempt to consider agrarian environments as a single analytical construct is a means as well of being alert to changes in the social identities of people who live in, and help comprise, these changing landscapes. The hybridity and plasticity of landscapes, when comprehended as something processual, leads to a consideration of the politics of identity and other similar processes through which social typologies are constructed, politicized, deployed, and unraveled. The reification of landscapes into the environmental or the agrarian, our perspective suggests, is closely allied to the politics of naming, to the fetishization of social difference. For instance, the use of the term "environment" to represent autonomous nature, divorced from the agrarian, also facilitates the use and fetishization of allied ideal-typical concepts such as "woman," "indigenous," "community," and "local."[9] A glance at some of the recent literature in environmental studies is instructive.

The past two decades have witnessed a fragmentation in writings about environmental questions. Although a significant literature continues to be produced on global environmental change, many scholars of the environment have also moved away from treating environmental problems as having a primarily global character.[10] Focusing on institutions, demography, and social identities, and their relationship with resource use patterns and environmental problems,[11] newer writings have begun to consider explicitly the interests of different social actors (Agarwal and Narain 1992). Much recent political-ecological research has documented how environmental degradation is embedded in exploitative relations between regions and nations.[12] These scholarly trends have helped to make the analyses of environmental issues more political.

The specific form of this shift toward the political has been a greater emphasis on the group identities and interests operating in formal and informal institutions that regulate the use of renewable resources. Underscoring the gendered nature of environmental degradation, for example, has done much to reveal that women often bear the brunt of environmental degradation and seldom have much voice in how degradative processes can be reversed or halted.[13] Similarly, nation-states today can be seen neither as constituting the sites of a monolithic rationality[14] nor as unproblematically representing the interests of heterogeneous communities within their borders.

The recognition that communities can be efficient resource managers and form viable alternatives to the contractual or hierarchical institutional arrangements embodied by markets or states[15] has been one of the central achievements of the large literature on commons and local communities. A spate of recent practical initiatives on forests, irrigation, wildlife, and pastures, where governments have attempted to devolve resource management responsibilities to communities, can be seen as a significant consequence of the realization that communities can be viable managers. The focus on community has also brought to the fore the possibilities inherent in a series of concepts related to community, chief among them being the local and the indigenous. The significance of local-level processes and the conflicts between localities and larger interests have become a staple of research on the environment.[16] Indigenous knowledge, at the same time, has emerged as the foundation of a range of programs concerned with the environment. Current writings point to the relevance of indigenous knowledge to processes of development and conservation and highlight the role of indigenous peoples in environmental management,[17] especially in contrast to "outsiders," who are often seen to possess little stake, and therefore little interest, in the stewardship of resources.

The emphasis on specific social identities has helped shift attention toward more marginal social groups and led to a greater appreciation of the differentiated impacts of environmental problems. Unfortunately, it has also congealed a series of dichotomies that threaten to become naturalized. "Woman," "indigenous," "community," and "local" have become central as building blocks for specific streams of environmental writings. The populist potential of these constructs is enhanced by using them in opposition to others such as "man," "Western/scientific," "state," and "global."[18] For some ecofeminists, women as the embodiment of nature are also its spontaneous guardians and protectors. The projected unity is shattered only because of the intrusions by a patriarchal society. Indigenous knowledge and people, in an analogous fashion, have come to symbolize the aspirations of those wishing to return to an earlier, less complicated, more ecological state of existence. As the holding place for stewardship possibilities, indigeneity may seem politically appealing to many. Similarly, the reification of women and the indigenous may serve an immediate political cause. But the easy solace of such reifications can well sacrifice long-term gains for all classes of people whose varied experiences are concealed by hastily composed social categorizations. They close off possibilities for alliances across politically and analytically expedient groupings, which for that reason might appear unsettling. Such alliances must nonetheless be explored if politically marginalized groups are to claim a share in power.[19]

The emergence of community as a source of hope for those who believe markets to be incapable of privatizing externalities and who are disappointed with the achievements of centralizing states has led to mythic visions of place-based communities. Through the mechanism of such mythical communities, analysts can unite the diverse social goals of equity, sustainability, and development. These visions hinge on a naturalized past where communities protected the environment and lived as one with it. But such versions of community, and of the ability of communities to manage local resources, must always remain troubled by research that points to internal stratification and oppression within communities, and the politically asymmetrical position of all communities attempting to behave as autonomous actors ranged against powerful political or economic interests.[20]

The complex political and analytical moves that have conjoined the local with community have also meant that the local has assumed oppositional overtones against the homogenizing influences of a global modernity. As the presumed site of cultural diversity, isolable local spaces may often be seen to constitute gaps and possible sources of autonomy in a world rapidly being pushed to experience a blandly uniform consistency. But such visions of the local and the local community cannot address any acute questions about the nature of the homogenizing influences modernity is supposed to introduce, or about the existence of breaches and fractures within modernity.

It may not, then, be surprising that nature itself is symbolized by the sequestered natural park within which biodiversity in its various forms can be preserved. But this way of conceptualizing and operationalizing the protection of nature ignores a tremendous array of historical ecological evidence that demonstrates the multiplicity of strategies through which humans have been instrumental in producing nature even in what are assumed to be the remotest and most virginal landscapes. Environments have histories from which humans cannot be excluded (Sponsel et al. 1996).

The attempt to identify women, the indigenous, communities, or the local with the natural environment is attractive to many scholars, activists, and policymakers alike. The appealing aspects of such identifications cannot be denied, and their utility for drumming up support is evident in the passions created around them. But we must also recognize that such attempts often reduce complicated social and historical dynamics and the fraught nature of social identities to mere caricatures. The resulting simplification and reification of categories not only flattens the complexity of phenomena that are thus imagined but also limits the possibility of enriching the study of environmental politics with new theoretical insights.

In contrast to the naturalized "environment," agrarian environments are places that can neither be isolated as parks nor be seen as the obvious centers of Vavilovian biodiversity. They are local spaces, as all experienced space perhaps is. They are home, however, not to the striking and exoticized indigene or the essentialized natural woman but to complex social formations and identities that reflect the diversity and flux of their landscapes. The communities that live in them are not the self-sufficient and harmonious formations currently the darling of many conservationists. Rather, these communities are unavoidably fragmented politically and are located and shaped in wider sociopolitical contexts toward whose construction they contribute. These are the convictions that inspire this volume and provide the basis for a schematic review of the literature on environmental studies in south Asia.

THE STATE OF PLAY IN SOUTH ASIA

Two recent and important collections of essays on the environmental history of south Asia have attempted to set the terms of debate and define a research program for future work on environment-related scholarship in India.[21] The introductions to both the volumes survey the development of the fledgling field of environmental history in south Asia and conclude that the field was deeply influenced by the concerns of the Indian environmental movement as it emerged after the 1970s. We are powerfully reminded that environmental history in India was inspired by a radical critique of government and development that was building up amid the Sarvodaya movement and other anticentral government sentiment of the early 1970s. We also observe a clear link between environmental history, international Green politics, and the sharpening of anxieties about tropical deforestation, land degradation, and its relationship to global futures.

The agenda that has been set, and begun to be accomplished, by Arnold, Gadgil, Grove, and Guha has brought environmental concerns to the fore in older social and economic history debates in Indian studies while charting much unexplored territory in the historiography of colonial India. Among other things, they have started to trace the chronological course of ecological sciences, offered materialist and culturalist histories of the nature-culture relationship, initiated studies of the environmental impact of urbanization and technological transformations in agriculture, analyzed environmental degradation as it affects specific resources such as water and forests, signaled the role of modern state formation in resource exploitation, and indicated how science, technology, medicine, and law can be studied as colonizing projects.[22] The writings of Grove in particular, and others in his wake, have detailed the development of colonial discourses about nature, science, risk, and control of natural resources; suggested that forest history is the legitimate focus of environmental history against those who have focused on the urban or the arable; and posited differences between indigenous and colonial constructions of nature and their interaction in specific settings.[23] Writings about the environment in the Indian context have thus opened up many new, exciting vistas. But they have also tended to accept the tenacious and obscuring dichotomies regarding the foundational concepts already identified. The acceptance of these dichotomies, driven in part by agendas oriented toward practical politics, is visible especially in the writings of Guha and Gadgil.[24] In explanation of these scholars, it must be said that their concerns were more to establish the environment as a legitimate domain of study. Indeed, without their contributions, it would be that much harder to call for studying agrarian environments, or to argue for dismantling the easy separations between nature and culture, indigenous and scientific, community and state, that are erected and enacted in the defense of disciplinary boundaries.

Some recent contributions to the study of the environment have begun the move toward recognizing the interpellated nature of the agrarian and the environmental. New research has begun to demonstrate the interlinked livelihoods of forest-dependent communities and local economies and the networks on which the livelihoods of tribal groups were dependent (A. Prasad 1998). Dangwal (1998) documents how migration from the hills was linked to a whole series of demographic, agricultural, and land use changes that can be ill understood if we remain locked within a concern with either the purely agrarian or simply the environmental. Building on similar ideas of the links between the natural and the cultural, Rangarajan (1998) suggestively argues that views about the animal world, about the dangers or beauty inherent in wildness, are closely affiliated with forms of land use, rhythms of agrarian expansion, and social relations among humans around production processes. These existing arguments help found the grounds for a third generation of analyses of environmental processes and politics. The chapters of this volume extend the limits of these arguments by undermining the conceptual and social identities consolidated by separating the environmental from the agrarian, and elaborating on the politics of lived experiences.

The chapters of this volume suggest that the relations between identities and interests articulated during resource allocation conflicts are contingent on specific ecological, historical, and cultural contexts. Further, unitary subject positions entailed in dichotomous analyses always have a repressive as well as a coalescing function. The idioms of the local, or the indigenous, or the community, opposed to the global, the outsider, or the state/market, can prove fruitful for rallying support. But support for these causes, as they have hitherto been constructed, also simultaneously requires that internal differences be glossed over and erased. The implicit larger argument we present is that there are no conceptual categories that endure in the same way or mean the same thing to all people. Especially when they take the form of politically charged binaries, categories can assume a phenomenological life of their own, but that process itself needs to become an arena of inquiry.

A second intervention of this volume is the articulation of the concept of agrarian environments. Despite pertinent distinctions drawn by south Asian and other scholars between developed and developing country environmentalisms,[25] both these modes of analysis and political mobilization share an approach that treats nature, its degradation, and its conservation by separating them from the world of rural production.[26] The separation between the arable and the nonarable in the rural environment is consolidated by focusing on questions of deforestation, shifting cultivation, hunting and wildlife, and the degradation of pastures from the perspective of state policy and peasant resistance. This delinking of nonagricultural and shifting agricultural livelihoods from the settled agrarian economy has ironic consequences. Environmental historians of India are quick to chide agrarian historians and rural sociologists for neglecting the nonarable world that surrounds the arable world they study. But the environmental historians themselves only insufficiently examine the role of agrarian change in the emerging patterns of environmental transformations and conflict.

Even when environmental problems are located in a specified relation to the agrarian economy of intensive crop production, the focus has usually been on one of two things: first, on the problems of technological change that are somehow considered superior to power, or to economic and social relations of production;[27] second, on unilinear accounts of deforestation and consequent agrarian distress, following the directions charted by the large empirical and statistical undertaking of the Duke University project.[28] The connections that have been established by recent work have thus tended to be unidirectional, explaining environmental decline in terms of new agricultural technologies, or blaming famine and other rural hardships on deforestation.

Let us take the case of famines. A range of writing in agrarian history has described famines, their social and economic consequences, and, in passing, their relationship to peasant unrest. There is little scholarship, however, on the frequency, intensity, spread, and recurrence of famines, especially with a view to finding the connections between environmental change, the politics of environmental management, and agrarian relations.[29] Vinay Gidwani's contribution to this volume raises some of the issues pertaining to the material complexities of famine and its relationship to environmental history. Cultural-geographic studies of the construction of India as famine prone, in the late colonial period, are the subject of Darren Zook's chapter. "How did Indian landscapes come to be imagined by different groups as famine ridden in the nineteenth century?" he asks. His analysis of the representations of hunger in colonial and nationalist writings shows how these literary productions depict environmental crises as emerging from a profound lack in agrarian landscapes.

The chapters of this volume thus argue for a more nuanced and dialectical relationship between the world of agrarian production and environmental change. This allows us to point out, for instance, how narratives of deforestation are constructed culturally in competing representations of changing landscapes. Questions about the cultural construction of agrarian landscapes were notable omissions from the literature in U.S. or third world scholarship until recently. A spate of work promises to remedy this absence in Western scholarship.[30] Writings on the environment in Africa have also begun to consider the issue.[31] Environmental history in India, however, remains little influenced by these new approaches to studying landscapes as cultural representations of contests over resources and identities.[32]

A third problem in the environmental history of India relates to its historicism. Because of the opacity and flatness with which current histories of the environment treat concepts such as the state, traditional communities, and capitalism, their assumptions about the relations among these concepts become suspect. Their flawed historicism often assumes chronological and epochal divisions between the precolonial, colonial, and postcolonial periods that are not always useful. Instead, we need greater, more curious, and more insistent attention to the social construction of historical relations in environmental conflicts, how the past is inflected by current utopian aspirations, in short to explore the dynamic relationship between history and current moments. The recognition of how the past might be constituted by the present, or explorations of the ways in which the past provides legacies shaping regimes of management and contestation in the present, lift us out of conventional disciplinary domains and permit research that effectively integrates approaches and methods from the humanities and the social and natural sciences.[33]

THE CONTRIBUTIONS

We started our journey with two broad sets of themes and a conviction that they are related in potentially fascinating ways. One was the need to examine the cultural and political production of social identities in the context of environmental conflicts. The other was the desire to integrate agrarian and environmental politics. The integration of these two themes in the concept of agrarian environments allows a focus on the significant regional variations within modern India's agrarian environments, rendering the project of a south Asian environmental history a fundamentally comparative one. Environmental history and politics, we learn, have to be written at several levels of aggregation, the national or subcontinental being but one of them. South Asian agrarian environments need also to be placed in a wider universe of European expansion, international commerce in commodities, ideas, and images, and changing landscapes as one moves from west Asia toward southeast Asia.

Colonialism, ecofeminism, deforestation rhetoric, environmental catastrophism, green revolution technologies, international aid packages, and modern state formation are powerful elements in a developing international public sphere around environmental issues. The relationships of Indian agrarian environments to this sphere are changing in ways that begin to erase the moral distance between environmentalisms in first and third worlds. For instance, the rise of social justice movements in the United

States and Great Britain is symptomatic of the new salience of equity issues in environmental struggles worldwide. The attention to community involvement in resource management is burgeoning both in the West and in the South.

These fairly new connections between environmental politics across the developed-developing divide prompt us to examine the production of analytic categories and social aggregates. The chapters of this volume collectively demonstrate how categories such as gender, community, caste, state, technical knowledge, and binary land classification schemes are constructed and can be contested. At the same time, an assumed affective social cohesion can become—at least temporarily—an actual interest-based collective consciousness. Caste and gender are good examples of such transformative potential in gross social aggregations. Occupational markers, like pastoralist, or bureaucratic functional ranks, like extension officer, carry the same possibility.

The chapters are located in a rough chronological order. Without a necessary adherence to linear historical time, we nonetheless want to suggest that the themes we are marking in this introduction apply to writings about the environment in the Indian context both historically and contemporaneously. Rangan, Baker, Saberwal, Springer, and Zook, in particular, examine the social relations embodied in the state to expose its shifting priorities, internal fractures, and modes of representation. Guha's chapter marks a shift in the location of analysis from the state and uses historical evidence to penetratingly examine the idea of community. The following three chapters are located in the present, but they are equally alive to the issues that raise their heads once the easy identity between community and conservation is questioned.

In earlier analyses, agrarian historians had certainly explored such fractures within the community and the state, pointing to the politics to which both these conceptual formations are home. These historians' analyses had focused on the differentiations within the state apparatus and on changes over time in the factors and perceptions that shaped state policy. But in carving out the environment as a separate domain of study, existing scholarship seems, ironically, to have lost that insight. In this sense, some of the arguments in the following chapters are only an attempt to regain that nuanced understanding of state and community that agrarian historians of south Asia have always had. Of course, we should add that the separation of the chapters into those focusing on the concept of the state and those looking more closely at community is not an attempt to indicate that community and state are somehow separate from each other. Certainly the arguments in the chapters do not support such an improbable stance. Instead, they go a long way toward suggesting that the contours of social formations called the "state" and the "community" are often constructed interdependently. The existing organization of the chapters follows a particular social scientific approach to state/community but also reflexively questions the foundations of that approach.

Rangan's chapter is part of a new scholarship that is recovering for the study of the environment the same sophistication that characterizes the treatment of the state by agrarian scholars. She discusses the different phases in the government's attitude toward, and interest in, forested land in the Uttarakhand from 1817 to 1947. Her investigation seriously undermines the belief that during this entire period a colonial state worked to appropriate forest resources in the name of a coherent scientific forestry policy. In pointing to the different phases of colonial forestry policy in Uttarakhand, and the motivations that led to these phases, she connects the history of forestry to larger colonial concerns. She thus helps undermine the boundaries placed around constructions of environmental histories and shows how these histories and politics develop in relation to much wider concerns.

If Rangan points to the historically changing nature and interests of the colonial state, Baker, Saberwal, and Springer employ different strategies. Baker describes how the belief in the opposed interests of the state and the community is untenable. The colonial state in Kangra, according to his account, was simultaneously facilitating efforts of communities to expand agriculture and restricting their interests in forests. The same desire to increase revenues led to very different kinds of negotiations between the state and the community depending on the nature of resources, extractive institutions, and technologies. At the same time, the imposition in the hills of categories of property from the Indo-Gangetic plains created forms of communal organization of land that today are often seen as belonging to a distant, traditional past.

Saberwal adopts a more direct route to demonstrate the internal fractures within state institutional structures and an overall incoherence in state objectives. He does so by reporting on the contradictory understandings displayed by the revenue and the forest departments in the matter of defining and combating a problem termed "overgrazing." The exacerbation of differences between these two rural land management arms of the government generated and fed on a desiccation discourse that gained ever greater influence on forest policy. The continued vitality of the discourse can be seen today in how it permits the forest department to extend control over land. Saberwal's arguments are reminiscent of other work that examines bureaucratic conflicts in the colonial period, especially between the land revenue and the forest departments.

Springer's chapter turns attention to the internal fractures and permeability of the contemporary postcolonial state. By showing how agricultural field agents in Tamil Nadu are both transmitters and objects of development, she reveals their multiple locations within state hierarchies and social formations. Their subject positions and relationships with farmers serve to transform, and at the same time extend, the aims of development by rendering them in terms that are locally meaningful.

Several of these analyses of the state also point to the politics of representation—a theme taken up explicitly by Zook in his study of images of famine and hunger in south India. Zook shows how images of India as a land and Indians as a people who were constantly beset by famine and hunger were used by colonial officials as well as their Indian critics. Although there was considerable variance in the objectives and intentions of various groups in seizing upon famine and hunger as problems afflicting Indians, their strategies led to similar consequences: the enshrining of representations that depicted India as incapable of dealing with these problems. His contribution goes on to discuss the reasons behind the enduring power of such representations of poverty and famine today.

The next set of four chapters by Guha, Jackson and Chattopadhyay, Gururani, and Robbins takes up for investigation perhaps the most powerful locus of imaginings in current environmental writings: community. Guha turns his critical attention to the belief that autonomous premodern village communities were the repositories of a conservationist ethic. He points to the existence of numerous conflicts over land, pastures, and forests, and he shows the negotiations through which these resources were used and often appropriated by the more powerful strata in premodern agrarian society.

His study shows the value of placing the discussion of ecological change back in agrarian environments to understand the frictions within, and the construction of, community.

Community emerges as a complex and conflict-ridden world in seventeenth-century western India in Guha's work. That characteristic of community, its tendency to disappear when the relations of its members are examined closely, has not changed today. Robbins's detailed study of pastoralists in Rajasthan shows how the same ecological changes differentially affect members of pastoralist castes. Opening the category of pastoralist, he shows how internal stratification allies the interests of some pastoralists with landowning elites in a village and of others with more marginal agricultural producers. His analysis problematizes images of the village as a coherent, bounded community, and that of migrant pastoralism as an undifferentiated subsistence activity.

Jackson and Chattopadhyay's work on a single village in Jharkhand similarly shows how caste identities critically shape access to forest resources. Questioning the often taken-for-granted category "woman," they explode the myth that women have any necessarily unified consciousness about resource use or management. But theirs is not simply a deconstructive exercise. They also show the specific social and economic forces that produce particular forms of consciousness about resource access and use. In showing the internal politics and dynamics of village communities around issues of caste and gender, their chapter moves environmental accounts toward a needed emphasis on how communities themselves are produced or undermined in struggles.

Gururani, like Jackson and Chattopadhyay, uses caste to problematize analyses of forest access that have so far used simple gendered approaches. Her work on practices that women follow in harvesting forest products in different villages in the Kumaon Himalaya shows the ways in which the category of "woman" is internally fractured and under constant renegotiation and contestation by the very subjects the category presumes to represent.

Of all the chapters in this volume, Gidwani's description of the changing agrarian landscape in Matar Taluka in Gujarat and its relationships to the environmental context addresses the reciprocal relationship between agriculture and the environment most frontally. He accomplishes this by examining the validity of a widely accepted polarization thesis that presumes to explain the emergence and consolidation of social inequalities in agrarian societies. The caste communities on which he focuses to lay bare the limitations of existing arguments about the generation of inequalities, he argues, emerge in their concrete relationships with what is glossed as nature. However, nature itself is neither "an autonomous entity, nor a mere imagining." The argument holds for all communities that encounter nature through some sort of labor. Gidwani's chapter would be valuable enough for just this insight. But he goes several steps beyond. In charting four "agrarian environmental" mechanisms through which agrarian and social change unfolds in Matar Taluka, he provides the beginnings of a framework that potentially is relevant to any analysis of agrarian change. What we have in his chapter is a carefully and insightfully elaborated explanation of the means through which agrarian environments come into being, both as social formations and as identity-related transformations.

The second part of the volume, entitled "Reflections," presents the thoughts of two scholars on the themes and theses explored in the different papers. Using their extensive research experience in the subcontinent, David Ludden and Ajay Skaria bring different theoretical perspectives to bear on the original research presented in this collection. Ludden provides an appropriately broad sketch of the nature of agrarian environments in a larger perspective, covering the entire subcontinent, and examining as well some of the issues related to urban development. Skaria focuses on the conceptual and theoretical relationship between modernity and environments, especially to discuss how the colonial or postcolonial variants of modernity are integrally connected in the very production of the idea of nature or the natural. Their comments, together with the earlier chapters, show the dynamic interplay between empirical research and theory construction.

The discussions in all the chapters in the volume demonstrate the larger implications of questioning scholarly practices that divide environmental relations along conveniently bifurcating axes, of investigating the agrarian intersections of environmental conflicts, and of focusing on the "field" as well as the "archive." The volume provides a new emphasis on politics and the differential impacts of environmental policies by focusing on constituent groups of the so-called community, and categories such as "indigenous" or "woman." Further, by indicating how many of the foundational concepts in the discourse of environmental studies are themselves constituted, the chapters in this volume force a more nuanced appreciation of these concepts. In providing a more textured analysis of how, for example, categories such as "woman," or the "community," or the "state," emerge, the chapters hint at possible alliances across naturalized divisions between the community and the state, or across gender lines, or in relation to indigenous and outsider populations. Exploration of such potential alliances may be seen as a survey of dangerous grounds—co-optation, after all, is always a possibility that lurks in the wings of such a venture. But it can also prove critical to environmental initiatives by pointing to ways in which existing divisions can be breached. Finally, insisting on the historical and contextual specificity of environmental conflicts also implies a refusal to bow too easily to the seduction of theoretical generalization, a seduction that has been responsible for numerous painful, wrongheaded, and unsuccessful policies for environmental protection and conservation.

NOTES

1 The point is especially applicable in relation to studies of the environment in south Asia, but also relevant more generally. See, for example, Rolston 1988.

2 In this context, work by scholars such as Buttel (1996) and earlier by Redclift and Benton (1994) has usefully pointed to relationships between the social and the environmental/natural. See also Cronon 1995.

3 The normative assessment of this originary, imaginary state of nature differ, of course, across the works of these political philosophers.

4 This image of a pristine environment is analytically analogous to the "state of nature" in the thought experiments of political theorists such as Hobbes and Rawls. In much environmentalist discourse, however, the idea of a "pristine environment" assumes the status of something that actually existed in a fairly recent past, and which is now irrevocably gone.

5 These important revisions of an otherwise flawed historiography are outstandingly represented by Bayly (1983) and Washbrook (1988) and well discussed by Stein (1989), but they worked to strengthen a scholarly distinction between agrarian landscapes and other nondescript ones.

6 For more on this, see S. Guha 1992, 3–6.

7 As Bhattacharya, in his introduction to a special issue of *Studies in History* on "Forests, Fields, and Pastures," argues, "If the old agrarian history neglected the forests and pastures, environmental history now has banished the peasant fields and farms from the realm of historical concern" (1998, 165).

8 See the contributions of Partha Chatterjee and Asok Sen to *Subaltern Studies I–V* (Chatterjee 1982, 1983, 1984; A. Sen 1987). A limited exception to the failure of elite and subaltern studies alike to study the interrelations between agrarian and environmental issues might be the work of Eric Stokes (1978) and David Washbrook (1978).

9 For the most part, after this initial indication of our deep disquiet with the use of these naturalized terms, we do not use quotation marks each time we use one of these terms. Given that the objective of the volume is precisely to examine the formation and deployment of these problematic categories, we trust the reader will take our discomfort with these terms from the context of the discussion.

10 But for a relatively recent example of the obsession with global environmental impacts, see Shah 1998.

11 Several of the essays in Arizpe, Stone, and Major 1994 are serious efforts to link institutional, demographic, and environmental variables. Agrawal and Yadama (1997) provide an example from the Kumaon.

12 Notable works are Blaikie 1985; R. Guha 1989b; Peluso 1992; and Schmink and Wood 1992. These monographs have helped to define the emerging field of political ecology. For studies tracing the origins of political ecology to different environmentalisms, see A. Atkinson 1991 and Eckersley 1992. More recently some scholars have defined the scope offered by political ecology for social scientific research on environmental questions. Useful essays in this genre are Neumann 1992; Bryant 1992; and Peet and Watts 1996.

13 See B. Agarwal 1994; Fernandes and Menon 1987; and Merchant 1980.

14 See, for example, T. Mitchell 1991 and Gupta 1995.

15 The definitive work in this regard is Ostrom 1990. See also Berkes 1989; Bromley 1992; McCay and Acheson 1987; and Wade 1988.

16 See, for example, the literature review in the research document produced by the IFRI (International Forestry Resources and Institutions) Research Program (IFRI 1993). For a sample of new political ecology research that illustrates these approaches, see various articles in the collection "Voices from the Commons: Evolving Relations of Property and Management," *Cultural Survival Quarterly* 20, no. 1 (1996).

17 For the role of indigenous knowledge in development and conservation, see Warren, Slikkerveer, and Brokensha 1995. For writings about the role of indigenous peoples in environmental protection, see Greaves 1994 and Brush and Stabinsky 1996.

18 But see Agrawal 1995 for an interrogation of the concept of the "indigenous." See also Ilahiane 1993, which focuses on the divisions within a community on the basis of religion, class, and ethnicity, and how they lead to exploitative relationships among community members. Sivaramakrishnan 1998 discusses the exclusions and coalitions through which "community" is constructed.

19 This point is particularly important for discussions related to the environment because social movements around the environment are generated by irreducibly plural constituencies. The consideration of environmental and agrarian problems calls for analyses that do not use class, gender, ethnicity, or caste as the only lens for viewing. Writings on the environment have increasingly begun to pay attention to this issue (Buttel 1992; Mellor 1996).

20 Insightful discussions of how community is constructed and deployed are present in Li 1996 and Moore (1998). For a general review, see Agrawal 1997.

21 Arnold and Guha 1995; Grove, Damodaran, and Sangwan 1998. See also Ramachandra Guha 1993.

22 Ramachandra Guha 1989b; Gadgil and Guha 1992; D. Arnold 1993; Rangarajan 1994.

23 Grove 1995; MacLeod and Kumar 1995; Rajan 1994.

24 We may, however, argue that Grove's seminal work on environment and forestry has also contributed to some extent to institutionalizing the separation between the agrarian and the environmental.

25 See, for instance, Ramachandra Guha 1989b; and Guha and Martinez-Alier 1997.

26 Some exceptions to this assertion would be Worster 1979; Blaikie and Brookfield 1987; Beinart 1984; D. Anderson 1984; Richards, Haynes, and Hagen 1985; Richards and Hagen 1987; Richards and McAlpin 1983; R. Tucker 1989; and Richards and Flint 1990.

27 Classic examples are the work of acknowledged pioneers in environmental history in India and the United States, respectively. See Whitcombe 1972 and Cronon 1991. The enduring influence of a technology-driven analysis of agrarian change and its environmental consequences may be seen even in recent work in India. See, for example, Agnihotri 1996.

28 Recent examples of such work may be found in Damodaran 1995 and M. Mann 1995.

29 See Dhanagare 1983 and the collection of essays in Hardiman 1993.

30 See Cosgrove and Daniels 1988; Schama 1995; W. Mitchell 1994; Mackenzie 1995; Harrison 1992; Hirsch and O'Hanlon 1995.

31 See, for instance, several articles in Beinart 1989; Fairhead and Leach 1996; McCann 1995; and Wilson 1995.

32 D. Arnold (1996) provides a pointer in this direction.

33 Further discussion of the ways in which we can integrate historical and contemporary approaches to environmental topics in India may be found in Sivaramakrishnan 1995. Ludden (1992) suggests a comparable reformulation of development's histories. For a discussion of history as politicized memory, see also Peel 1995.

Haripriya Rangan

State Economic Policies and Changing Regional Landscapes in the Uttarakhand Himalaya, 1818–1947

Narratives of environmental change in the Uttarakhand Himalaya generally fall into a declensionist genre, invoking images of a pristine and isolated region located in the eternal past until the advent of colonialism and capitalist development marked the beginning of relentless ecological degradation (see, for example, Bahuguna 1982; Berreman 1989; Bhatt 1987; Dogra 1983; Gadgil and Guha 1993; Ramachandra Guha 1989b; Shiva and Bandyopadhyay 1986a, 1986b; Weber 1988). The dramatic power of these narratives is enhanced by attributing a particularly malevolent role to the colonial state. The colonial state is caricatured as overwhelmingly powerful, autonomous from and thriving on antagonistic relations with civil society, and single-minded in its predatory pursuits that inevitably cause ecological degradation and impoverishment of Himalayan communities (exceptions to this view are R. Tucker 1983, 1991; J. Richards 1987; Chetan Singh 1991). The colonial state plays the role of villain in the relentless Manichaean struggle of environmental change. It is the destroyer of precolonial harmony, the promoter of modernity against hallowed tradition, the harbinger of Western patriarchal modes of capital accumulation that undermine "Oriental" feminine principles of nature, the diabolical agent of capitalism that transforms ecological utopias into lifeless terrains. The environmentalist Vandana Shiva, for instance, uses this imagery to eulogize the passing of simple and self-sufficient peasant lifeways organized around the "feminine principle of nature" as they encountered "western patriarchy," and "maldevelopment" promoted by the colonial and postcolonial state (Shiva and Bandyopadhyay 1986a, 1986b; Shiva 1989a, 1991; Shiva and Mies 1993).[1] Social historian Ramachandra Guha employs parallel tropes in his description of peasant resistance in the Indian Himalaya, where, according to him, local communities routinely rose up to defend their "moral economy" based on "ancient community solidarities and sets of values" against the depredations of colonialism, capitalism, and "scientific forestry" (Gadgil and Guha 1993; Ramachandra Guha 1989b, 21, 48–63).

Stylized representations of the colonial state are key in transforming narratives of environmental change into powerful and compelling myths. Such assumptions regarding the nature of the colonial state—that it is a monolithic entity and single-minded in its predation of civil society—logically lead to two conclusions: first, that colonial rule was based on a remarkably coherent and tightly orchestrated set of policies that remained unaltered by the forces of necessity or contingency; second, that colonial administrators were endowed with extraordinary capabilities that would normally fall within the realm of demonic power or divine omnipotence. Both implications are historically inaccurate and frankly implausible. However much they may appeal to nationalist sentiments, such representations of colonial rule are extremely unhelpful for understanding the processes of ecological change in regions. They inhibit the possibility of understanding the history of institutional actions, changes in global and regional economies, and social practices that have, over time, reworked the differentiated social and ecological landscape of the Indian Himalaya.

This chapter has two aims: first, to set out an alternative analytical framework that focuses on *processes of governance* rather than on the presumed inherent and immutable character of precolonial or colonial states; second, to provide an account of ecological change in the Uttarakhand Himalaya between 1818 and 1947, a period spanning both pre-British and British control. To meet these aims I analyze: (1) the political and economic processes at the global and regional levels that created pressures for state intervention, (2) competing demands that shaped the forms of state intervention, and (3) conflicts, disputes, and negotiations that redefined the exercise of control and governance by state institutions. The chapter shows that ecological transformations in Uttarakhand during British rule reveal a complex landscape repeatedly inscribed and incompletely erased by social actions emerging from the interplay of these three processes.

What the state *is,* at any point in time, is shaped by what it *does;* the institutional form and functions of state are, in this sense, mutually constitutive. As Peter Evans points out, Weber's definition of states as "compulsory associations claiming control over territories and the people within them" does not reduce the complexities of analyzing what states do (Evans 1995, 5–6). Weber's definition hinges on "claiming control," and thus on the functions and processes by which states direct social action within their territories. These functions and processes, in turn, produce forms of rule and governance that are both differentiated and reflective of the constantly changing relations between administrative institutions and other institutional practices (for example, in markets and everyday life) within particular territorial boundaries.

Over the past two centuries, states, colonial or otherwise, have not only performed their conventional roles of war making and maintaining internal order but also attempted, in varying degrees, both to foster economic growth and to ensure a modicum of social welfare within their territories. Their modes of government, which Foucault calls "governmentality," have been shaped by how these aims could be achieved, alongside more conventional functions, through their citizens and subjects (1991, 87–104). Foucault describes the actual practices of governance of the modern (nation) state-in-the-making as the "daemonic coupling" of the "city game" and "shepherd game." This refers to the making of a form of secular political pastorate that couples the "individualization" of citizenship (uniformity in treatment of individuals by law) with "totalization" of subjects (caring for each member of the territorial community, or "flock"; see Gordon 1991, 8).

Viewed from this perspective, therefore, the processes of governance by states have, with varying degrees of skill and success, involved activities that include not only lawmaking and law enforcement but also the making of distinctive territorial communities through interventions aimed at strengthening the material welfare of the citizen-pastorate.[2] One could argue that interlinkages between world economies—which were forged in varying ways for more than nine centuries—and continuing processes of integration into a global economy were shaped by the involvement of city- and nation-states that have engaged in economic transformation within their territories (Abu-Lughod 1989; Arrighi 1994; Braudel 1977, 1982, 1984;

State Economic Policies and Changing Regional Landscapes 25

Ghosh 1995; Polanyi 1944). What this implies is that the processes of governance by states were, at any moment in history, not only influenced by social groups within their territorial jurisdictions but also shaped by changing relations with, and conditions of, these world or global economies. The aspirations, strategies, and projects developed by states in response to constraints and conditions prevailing in particular periods—which I shall call *dominant policy phases*—were often radically altered when they confronted new situations or problems emerging from these interactions, and also from the unforeseeable outcomes of earlier policies (see Foucault 1991; Gordon 1991). Indeed, more often than not, many ambitious policies and well-intentioned projects promoted by states within their territories were undermined by contingent outcomes of failures and even successes in other areas of policy intervention.

It is necessary, therefore, to recognize two facts about states, past or present: first, despite the analytically convenient distinctions made between state, market, and civil society, states have never been completely autonomous from the structures of everyday life, nor have they been mere adjuncts to market processes; second, their policies and modes of governance, or "governmentality," have been constantly shaped by conditions and processes both within and beyond their territorial control. Both factors (i.e., being enmeshed in social institutions within territorial boundaries, and being interlinked with economic processes beyond their territorial control) have constantly shaped the modes of governance pursued by states in complex and contradictory ways.

The evolution of British rule and its governance in different regions of India needs to be analyzed from this perspective. Clearly there is little evidence to support the a priori assumption that from its moment of entry into the Indian subcontinent, the motives of the British East India Company were to establish a gigantic centralized state administrative structure aimed at complete exploitation and subjugation of its native subjects. Even though the East India Company was established in 1600 to serve the interests of both English merchants and the British state, it did not continue to share the same purpose or interests throughout the eighteenth and nineteenth centuries. As the Company's role expanded to assume greater economic power and territorial control over the Indian subcontinent, its institutional policies often clashed with the interests of the Home Government and with the dominant interests of the newly emerging bourgeoisie in Britain. On the other hand, the Company's role in commerce and governance also benefited some social groups and regions over others within the subcontinent. Throughout these two centuries, the Company was constantly contesting and evading attempts by the British Parliament to limit its powers at home and abroad. The Company challenged its critics at home, dealt with its European competitors in India, altered its motivations and strategies as new economic opportunities appeared on the horizon, and forged new institutional roles, linkages, and relations between different rulers, social groups, and regions in the subcontinent. For more than a hundred years beginning in 1757, the British East India Company scrambled through these opportunities, conjunctures, and constraints and in the process mutated from a commercial institution into a powerful Anglo-Indian monarch ruling over much of the Indian subcontinent. Direct control of India by the British Crown—that is, colonial rule—was achieved only in 1858, after years of protracted struggle between the British Parliament and the East India Company.

Historical analysis of agrarian and environmental transformations in the Uttarakhand Himalaya (or, for that matter, any other region in India) must explore the policies and strategies adopted by successive rulers—pre-British, the East India Company, semiautonomous rajas, the Anglo-Indian colonial government, the British-Imperial administration—as they attempted to maintain competitive advantage in the global and regional economies and contend with changing political relations within their territorial jurisdictions. The following sections focus on the period between 1800 and 1947 for eliciting the broad patterns of state control and policy making that shaped the ecological landscape of Uttarakhand before India's independence from British domination.

THE REGIONAL ECONOMY AND LANDSCAPE BEFORE 1800: THE PRINCIPLE OF TERRITORIAL ADVANTAGE

Uttarakhand comprises eight districts located to the northwest of the state of Uttar Pradesh. The region lies entirely within the Himalaya, sharing borders with Tibet to the north, and Nepal to the east. As part of the northern frontier of the Indian subcontinent, the territory has long been viewed as being endowed with immense geopolitical importance. For at least two hundred years before the region came under British rule, petty chieftains and kings in the region warred against each other to expand their control over the stretch of Himalayan territory that linked the Indian subcontinent with Tibet and Central Asia. The British East India Company was motivated by a similar intent when it gained control over the Garhwal and Kumaon Himalaya in 1815. Even at a later stage of colonial rule in India, the need to control the Himalayan regions bordering Central Asia and Tibet remained paramount for British administrators. Lord Curzon pronounced the Himalayan frontiers "indeed the razor's edge on which hang suspended the modern issues of war or peace, or life or death to nations. . . . The holders of mountains," he opined, had an "immense advantage against the occupants of the plains" (Woodman 1969, 7; also see Keay 1983). Himalayan territories were geographically strategic for rulers who attempted to control the mountain passes in the inner Himalaya so as to profit from the trade that moved between the Indo-Gangetic plains and Central Asia. The dominant principle of states in the region was to gain exclusive control over as many mountain passes and trans-Himalayan trade routes as lay within the region. Trans-Himalayan trade was the propulsive sector that sustained the regional economy, and agriculture and natural resource extraction remained important subsidiaries. The fortunes of agriculture and natural resource extraction were closely linked to that of transit trade; they prospered with growing trade and dwindled when it declined.

Sustaining a regional economy that primarily depended on transit trade was a delicate and challenging task for rulers of Garhwal and Kumaon, the two largest kingdoms in the western Himalaya. They faced constant threats of political instability from ambitious military administrators, war, and competition for transit trade from neighboring hill kingdoms. Their problems lay in the fact that, on the one hand, land revenues could not be assessed at higher rates for fear of desertion by cultivators to other areas where taxation was less burdensome; were this to happen, troops could not be maintained on dwindling revenues from land. On the other hand, any increase in transit duties and customs levied on trade would have a similar effect on traders, who would seek alternative passes through neighboring kingdoms where taxation was more lenient. The rulers of Garhwal, for instance, attempted to solve this problem by placing fertile areas and segments of trade routes under the authority of *faujdars* (military administrators), who collected revenues from villages to maintain armies that could be quickly mobilized in times of need. According to the revenue records examined by the British settlement officer in 1815, Garhwal's rulers generally drew about 70 percent of their revenues from the trade between Tibet and the Indo-Gangetic plains (Traill 1828; Atkinson 1882a, 289–91; Turner 1800);[3] only 30 percent of the total revenue was drawn from agriculture and went mainly toward supporting the king's armies (Atkinson 1882a; Rawat 1989; Saklani 1986; Walton 1910).

By the late eighteenth century, constant warring against invaders from the plains and between the two kingdoms led to a steady decline of both cultivators and traders in the region. When the Gurkha rulers of Nepal embarked on their ambitious attempt to consolidate their control over the entire stretch of the Himalaya, the Garhwali army had dwindled to no more than five thousand infantrymen, who were swiftly dispatched in battle (Rawat 1989; Saklani 1986). When the Gurkhas conquered Kumaon and entered Garhwal, the regional economy had already been weakened by the loss of trade.

THE REGIONAL ECONOMY UNDER NEPALESE RULE, 1804–1815

The Gurkha rulers of Nepal controlled Garhwal and Kumaon between 1804 and 1815, when they were defeated in battle by a military alliance involving the British East India Company and the king of Garhwal. The Gurkhas' ambition to control the entire stretch of the Himalaya had exacted a heavy toll on the Nepalese treasury. Thus the Gurkha rulers were essentially concerned with deriving as much revenue as possible within a short period, even if it meant stripping the region of every asset that could be extracted. In their attempt to recover the expenses of war, Gurkha administrators imposed heavy taxes on both trade and agriculture (Atkinson 1882a, 283–84; Walton 1910, 89), which further resulted in large-scale desertion of cultivators and dwindling of trade in the region. Households that could not meet their annual payments of produce or cash were sold as slaves and bonded servants, and it is estimated that nearly 200,000 people were sold at markets in the plains to meet the revenues demanded by the Nepalese rulers (Atkinson 1882a, 252–53; Walton 1910, 126–27). Entire villages were abandoned by communities; agricultural lands were reclaimed by jungle and wild animals.

The regional landscape of the Uttarakhand Himalaya mirrored the fortunes of rulers whose actions were shaped by the dominant principle of controlling trans-Himalayan trade. When transit trade flourished, fertile valleys and slopes near trade routes saw expansion of cultivation of rice, wheat, barley, buckwheat, and commercial crops such as amaranth, ginger, and turmeric. European travelers through the region commented on the deforestation along trade routes because of the growing volume of timber trade with the plains, and overgrazing by packhorses and goats (Chetan Singh 1991; Rangan 1995).[4] When trade declined because of war, villages were often abandoned, and jungle reclaimed areas that had previously been cultivated. The policies and actions of precolonial rulers in the region displayed no uniquely benevolent sensibilities—at least none that have been recorded—toward preserving the Himalayan environment or managing its natural resources.

CONTEXTS OF THE BRITISH EAST INDIA COMPANY'S POLICIES IN UTTARAKHAND

The British East India Company's acquisition of the Garhwal and Kumaon Himalaya was driven in part by the desire to profit from transit trade, similar to the principle employed by earlier rulers in the region. There was, however, an important difference: the East India Company was not a relatively independent ruler or a satrap paying occasional tribute to the Mughal emperor, but a commercial institution, whose functions and activities were sanctioned by a royal charter of the British Crown and regulated by the Company's accountability to its shareholders. The Company's activities in the Himalaya and the Indo-Gangetic plains were inextricably linked to its role in the rest of the subcontinent, as well as to the changing economic motivations, new social developments, and political alignments occurring in Britain and its competitors in Europe.

The British East India Company was a product of the "mercantile system," arising out of a relationship between kings and merchants that formed the foundation of the modern European state at a time when, as Wrigley describes:

> kings wanted money, which only merchants could supply, in order to expand their regal activities. Businessmen wanted public order, the freedom of operation within a large territory, which only a well-

financed royal administration could assure them. But soon they went beyond this basic aim and sought the positive help of the royal power in altering the supply conditions of labour in their favour and above all in securing advantages over their competitors in international commerce. "Mercantilism" thus implies that the economic unit is the state, that governments are in business, and that merchants are necessarily in politics. (1978, 21)

The Company was thus a mercantilist institution, carrying out an ambiguous mixture of commercial functions and political duties. However, its identity and functions grew more complex as it began expanding control over territories in the Indian subcontinent. The Company confronted its new seignorial role with a certain degree of confusion because on the one hand, it was a now legitimate Indian ruler,[5] but on the other, it was also expected to continue functioning as a commercial institution, accountable to its shareholders and the state in Britain. As an Indian monarch, the Company inherited the responsibility of providing appropriate policies and conditions for sustaining a prosperous economy for its subjects. This, among other factors, implied a healthy reserve of bullion in the treasury, and that revenues remained greater than expenditures. But as a mercantile institution competing with other similar European institutions, the Company was expected to function according to the policy of "staple"; this meant that it was primarily interested in moving goods, regardless of whether they were imports or exports, so that profits could accrue to the Company's shareholders and the state in Britain (Wrigley 1978, 23–26).

Economic historians have classified European mercantilist policies under three broad headings: *staple, protection,* and *provision.* The policy of staple was designed "to secure a bigger share of the profits of international commerce for one's own citizens." The policy of protection was inspired by fear of both overproduction at home and shortage of bullion and consequently aimed to reduce imports and magnify exports. The policy of provision aimed to secure the flow of imports, especially of "essential supplies," for the home country (Wrigley 1978).[6]

The three dominant policy phases of European mercantilism are useful for analyzing the changing strategies of the East India Company as it attempted to perform its hybrid role as both British merchant and Indian monarch in the subcontinent. They provide a framework for understanding why the Company embarked on particular economic ventures in its Himalayan territories during different phases of its rule. The key years marking the transition from one dominant policy phase to another during the Company's rule in the Indian subcontinent are 1813, 1833, and 1858: the policy of staple between 1813 and 1833 and the policy of protection between 1833 and 1858. The dominant policy phases under Crown rule can be identified as that of provision between 1858 and 1914, the beginning of World War I, and renewed protectionism and industrialization between 1918 and 1947.

In 1813 the British Parliament was pressured by free-trade proponents to deny renewal of the Company's charter, thereby ending its monopoly over the East Indian commercial trade. In 1833 the free-trade lobby in Britain was again successful in forcing Parliament to abolish the unfair advantages held by the Company; the Company's commercial operations were formally abolished, and the East Indian trade completely opened up to competition. From this moment onward, the Company functioned in a purely administrative capacity in India. In 1858, one year after the Indian Mutiny, the Company was abolished, and India was brought directly under the rule of the Crown. With the onset of World War I, colonial policy experienced yet another shift as the power of the Lancashire and free-trade lobbies weakened in Britain and the nationalist movement gained strength in the Indian subcontinent.

THE COMPANY'S POLICIES IN UTTARAKHAND: THE STAPLE PHASE, 1815–1833

Until its commercial functions were officially abolished by the British Parliament in 1833, the Company's administrators in India followed the policy of staple, deriving a large proportion of income from commerce. The Garhwal and Kumaon Himalaya were brought under Company rule in 1815, two years after its charter was denied renewal by the British Parliament. Northern Garhwal—comprising the present-day districts of Tehri and Uttarkashi—was returned to the raja of Tehri, and the remaining areas were retained under British control. Despite having borne the ravages of Gurkha rule, the region found favor in the eyes of its new rulers for various reasons, some of which emerged from the East India Company's rapidly changing political role in the Indo-Gangetic plains.[7] Its officers confidently viewed their territorial expansion as a reflection of their enormous power and assurance of continued prosperity in the subcontinent. Vast areas could now be controlled and coordinated for enhancing the Company's revenues from trade, thereby compensating for the losses incurred from being stripped of its commercial monopoly. By controlling Garhwal and Kumaon, the Company aimed to gradually gain control over the trans-Himalayan trade in *pashm* (cashmere wool), gold, borax, and salt by establishing trading posts in central Asia and Tibet. The Company's officers also viewed the region in terms of the geopolitical advantages it offered in the western Himalaya. Although full control of the western Himalaya was not achieved until after the Sikh War in the 1840s, the Company carefully scoured its northern horizons to anticipate potential threats to commercial and military competition from the eastwardly expanding Russian empire (Moorcroft and Trebeck 1841; Keay 1983, 17–34). The Garhwal and Kumaon Himalaya provided a strategic location for watching over the exchanges and transactions between imperial Russia and Tibet, as well as control of trans-Himalayan trade routes into Central Asia and the western outposts of the Chinese empire.

Between 1815 and 1833, the Company's officers attempted to encourage cultivators to return to villages that had been abandoned during Gurkha rule. They were keen, as previous rulers of Garhwal had been, to provide favorable conditions for the revival of trade through the region and to ensure political stability through settled agriculture. For the first three years, the Company continued to collect transit duties at the passes between Garhwal and Tibet, and between the hills and plains. But as restoration of agriculture and commerce showed very little progress, the settlement officer abolished all transit duties on trade in 1818, hoping the action would provide additional incentive for trans-Himalayan traders to carry their merchandise through Garhwal, as well as keep open the possibility for the Company's future entry into Central Asia. Owing to the scarcity of both cultivators and cultivable land in Garhwal, administrators offered land under relatively light assessments of revenue, encouraging people, wherever possible, to cultivate food crops or commodity crops such as hemp. The Company "procured a portion of its annual investment from the Garhwal and Kumaun hills in the shape of hemp," which meant that headmen of villages and principal cultivators in the region were given cash advances for cultivation and supplied the Company with hemp fiber in return. Hemp cultivation was seen as bearing excellent potential for the region, since the male plants yielded fiber for manufacturing rope, sackcloth, and hempen cloth, and the female plant yielded oilseeds and resin, or *charas,* which found ready markets in the plains (Atkinson 1882b, 799–802; Collectorate Records/Pre-Mutiny 1816–1857). During the first two decades of Company rule, cultivation expanded slowly, as did land revenue. The volume of trans-Himalayan traffic also increased as taxes on trade and transit duties were abolished (Atkinson 1882a, 289).

THE COMPANY IN UTTARAKHAND, PHASE 2: THE POLICY OF PROTECTION, 1833–1858

The Company had begun reorienting its policies in India before 1833, when its commercial functions were abolished by the British Parliament. By the 1820s, the Company's revenues from its territories began outstripping the earnings from international commerce, to the extent that it did not suffer a financial crisis when its participation in the East India trade was abolished (Stokes 1959, 37–39). Restricted to administration, the Company's role in India mutated from the cloven state of playing both British merchant and Indian ruler into a hybrid Anglo-Indian monarch retaining a strong mercantilist pedigree. The policies emerging from this hybrid role were largely protectionist, but different from the protectionism displayed by Britain in the early industrial period when it sought to promote the interests of its textile industry (Dewey 1978). The logic of mercantilist protectionism was geared toward conserving bullion, reducing imports, and increasing exports; it emphasized production rather than consumption, sale rather than purchase (Wrigley 1978, 24).

Between 1833 and 1858, the Company implemented its economic policies in two ways: first, it made infrastructure investments for increasing production of exportable agricultural commodities so that profits generated from their sale could be used to purchase essential imports and pay dividends to shareholders in Britain; second, the Company provided incentives for promoting ventures that would reduce expenditure of bullion for importing expensive commodities such as tea from China. These strategies complemented the Company's success in tenaciously holding on to its monopoly in the opium trade despite abolishment of its other monopolies. As the Company's officers expanded opium production in the subcontinent, they skillfully applied both free-trade rhetoric and imperial force to persuade the Chinese government to remove barriers constraining the free sale of opium and other commodities by British merchants at Chinese ports (Woodman 1969, 37–41).

The policy of protection led to the promotion of new economic ventures and projects in the Garhwal and Kumaon Himalaya. For the first time, perhaps, in the region's history, the policy emphasis was on export-oriented production rather than merely profit from transit trade. Trans-Himalayan trade continued to be encouraged but was now seen as complementing commodity production. Three major projects were launched in the region: commercial production of wheat and sugarcane, cultivation of tea, and construction of the Ganges Canal at the foothills of the Himalaya at Haridwar. The first of these centered on tea cultivation in Garhwal and Kumaon.

As far back as 1788, the East India Company had pondered the prospects of tea cultivation in India but did not pursue the idea with any vigor. Small-scale experiments for propagating Chinese tea plants were carried out at the Botanical Gardens in Saháranpur, a small town at the foothills of the Garhwal Himalaya. But the loss of commercial revenues and the high costs of importing tea from China gave the project requisite urgency. The discovery of indigenous species in Assam and Kumaon gave additional impetus to the prospects of cultivating tea in the Himalayan regions. Between 1835 and 1842, several nurseries were established in the region for propagating tea seeds and seedlings imported from China. As the tea plants began to thrive, artisans were brought over from China to establish manufactories for processing tea. Between 1844 and 1880, the area under tea cultivation in the Garhwal Himalaya expanded from 700 to 10,937 acres. The plantations—established and owned by the government—were centered around three sites in Garhwal, five sites in Kumaon, and one near the town of Dehra Dun.[8] As tea production increased and found favor with tea brokers in England, Dr. Jameson, the new superintendent of the Botanical Gardens at Saháranpur, advised further expansion of tea cultivation in Garhwal and Kumaon, observing that

a vast field for enterprise will be opened up, whether Government considered it worthy of their own attention, or it be brought about by private capital. Water carriage will soon it is hoped . . . also be a strong inducement, in addition to the above, to make capitalists invest their capital in this channel, and thus we trust ere long to see the hill provinces, which at present yield but a trifling sum to the revenues of the State, become as important, in an economical point of view, as any of those in the plains of Hindustan. (Dr. Jameson 1844, quoted in Atkinson 1882b, 895)

Although tea plantations in the northeastern Himalaya and Assam had begun shipping substantial quantities of the commodity from Calcutta to England, tea production in Garhwal and Kumaon was aimed at closer markets across the Himalaya in Tibet and Central Asia. In 1847 Dr. Jameson observed the vigorous purchase of Pouchong (black) tea by native merchants with undisguised pleasure, noting that the coarse Bohea tea produced in Garhwal was sold to Bhotias for export to Tibet, where it successfully competed against the tea imported from China. In his report on the foreign trade of the Northwestern Provinces and Oudh, Mr. J. B. Fuller remarked:

So far as the commercial interests of these provinces are concerned, the most interesting point in the traffic they transact with Tibet is the opening it might afford for the inferior classes of Kumaon tea, *which will not bear the cost of carriage to the sea board.* At present the markets of Tibet are closed by the united influence of the Chinese government and the Tibetan Lamas, who, having the monopoly of the wholesale and retail tea supply of the country, are naturally averse to the competition of a traffic in Indian tea, which might be more difficult to engross. . . . Yet the Tibetans on our frontier are compelled to purchase tea of atrocious quality, the price of which has been swelled by a long and difficult transport from the eastern extremity of the country; *while immediately across the frontier there are tea gardens whence they could be supplied with a better article, at a cheaper price, and with profit to the Kumaun tea planters as well as to the itinerant traders (Bhotias) through whose hands it would pass.* . . . So heavily is Kumaun tea handicapped by the expense of transport to Calcutta that the most profitable portion of the trade even now is that transacted in green teas with merchants from Central Asia who purchase the tea at the factory and carry it away themselves, saving the planters the expense and trouble of packing. (quoted in Atkinson 1882b, 901–2; Atkinson's italics)

Along with the expansion of tea cultivation in Garhwal and Kumaon, the government simultaneously launched a second venture that aimed at promoting the region as a "seat of European colonisation" (G. Williams 1874, 314–22), an idea that took hold when Europeans were legally permitted to own land in the Company's territories. The Company saw its own officers and administrators as potential entrepreneurs who would engage in agricultural commodity production for export and thereby strengthen its protectionist policies. The site chosen for promoting this venture in the Garhwal Himalaya was the Dehra Dun valley, seen as holding great potential for farming and horticulture. Although Gurkha rule had left Dehra Dun in disrepair and bereft of population as in the rest of the region, officers proposed the repair and extension of the ravaged canals that had once irrigated most of the cultivated areas. By 1837, the canals in Dehra Dun had been rebuilt and extended to irrigate large tracts of "uncultivated wastes," that is to say, any land that lay as abandoned fallows or jungle that had not been occupied by native cultivators. In 1838 nine land grants were given to European officers and merchants in the Dehra Dun valley, on the condition that land would be rent free for the first three years, after which rent would be charged at gradually rising rates until they reached the maximum rate at the tenth year. European entrepreneurs supervised the clearing of uncultivated wastes and planted them with Mauritius and Otahite sugarcane, cotton, wheat, and rice, looking forward eagerly to the substantial profits that would be derived from the sale of these commodities.[9] Land prices rose rapidly as European merchants and officers eagerly rushed in to buy remote tracts in the Company's territories, creating a land mania fanned by wild speculation and mysterious stories of windfall profits (R. Tucker 1983, 151).

The Company's policy of protection depended on the stability of agricultural production in its territories. Despite being a remarkably fertile region during the years of good monsoons, the Doab—the land between the Ganges and the Yamuna Rivers—suffered immensely during years of drought. The famine of 1837 to 1838 affected the entire Indo-Gangetic plain and spurred the construction of a large canal system for irrigating most of the Doab. In 1841, construction of the Ganges Canal began in Haridwar, an important pilgrimage site at the foothills of the Uttarakhand Himalaya, and extended across the Doab over the next twelve years.[10] By 1854, most of the canal was constructed and, according to the Canal Committee's estimates, was estimated to derive an annual return of nearly Rs 1.5 million (Cautley 1860, 28).[11] In less than two decades following the opening of the Ganges Canal in 1854, nearly 5,601 miles of main lines and distributaries had been constructed at a capital cost of nearly Rs 4.5 million. Between 1877 and 1878, the canal irrigated nearly 1.5 million acres in the western districts of the United Provinces. The expansion of irrigation brought about an increase in acreage under commodity crops, such as sugarcane, indigo, cotton, and opium, as well as wheat, an important export crop from the mid-1870s (Whitcombe 1972, 8).

The regional economy and landscapes of Uttarakhand underwent considerable transformation during the phase of protection. Tea plantations and agricultural commodity production led to increasing demands for labor for clearing lands, cultivation, and tea picking and processing. The construction of the Ganges Canal added to the demand for labor for excavation, lime burning, and brick molding and also increased the demand for timber and charcoal. Large quantities of wood were used for constructing buildings and roads, making charcoal for brick kilns, and curing tea. Timber extraction spread from along the sub-Himalayan tracts and into the lower ranges in Garhwal and Kumaon. A profitable timber trade emerged in the region as wealthier households involved in the transit trade redirected their efforts toward extracting timber from forests in higher elevations and selling it at high prices in the plains. Households with less access to capital continued to participate in the trans-Himalayan trade or sold smaller quantities of fuelwood and fodder at hill resorts and military cantonments. They also found seasonal employment in plantations, farming estates, and public works projects in the region. Despite the rapid transformation of the ecological landscape, the Company's policy of protection stimulated the regional economy of Garhwal and Kumaon. The period of prosperity for the region was unfortunately short-lived. The tea plantations and agricultural commodity production in the region were struck by a series of misfortunes and unforeseen outcomes of the Company's successes and failures with other policies.

The East India Company's projects in the subcontinent were often undermined by its detractors in the British Parliament. In 1841 Parliament passed a law banning British officers in India from private enterprise while still in the Company's employ. Despite the outrage expressed by the Company's officers, who pointed out that they had been given incentives and encouraged by the government to embark on entrepreneurial ventures, the British Parliament stood firm on its decision, and the Company was forced to comply. Production of agricultural commodities on farming estates in Garhwal and Kumaon abruptly ended. Machinery that had been imported for sugarcane milling lay idle as cultivation was abandoned. Grantees expected little or no compensation from the government and consequently stripped their lands of assets to recover as much of their investments as they could before selling their property at low prices to Indian merchants and landlords (G. Williams 1874, 324).

Tea cultivation faced problems of a different kind, arising not from the interference of the British Parliament but from the political tensions between Britain and China. Although the treaty signed in the aftermath of the Opium War of 1842 favored the Company's opium monopoly and British merchants aiming to establish trading posts along China's eastern seaboard (Woodman 1969, 37–41), it had the opposite effect for tea planters in Garhwal and Kumaon who were struggling to gain access to Central Asian markets. When the 1847 Boundary Commission established by the Company approached its Chinese counterpart for settling boundary demarcations between their territories and requested permission to establish trading posts in Central Asia and Tibet, they were denied both requests. In addition, the imperial government of Russia remained hostile toward the entry of British trade into its eastern provinces. Tea planters in Garhwal and Kumaon struggled well into the 1870s, steadily reducing labor and production on their estates, yet they could not muster the smallest profit. Atkinson complained:

> Trade with Central Asia, which at one time gave great hopes of proving remunerative, has been practically closed by the action of our Russian friends in putting a prohibitive duty on all articles imported

from India. The planters also complain that the reduction in duty on Chinese teas has also affected them injuriously. (1882a, 259)

Trans-Himalayan trade also suffered as cheaper imports of American borax became available, and as salt production increased in other parts of the subcontinent. Pashm wool was the only valuable commodity that survived competition, but the trading system had by this time also undergone substantial transformation. With the establishment of a woolen textile industry in Kánpúr and rising prices for wool, Tibetan traders sought to maximize their profits by directly supplying factories and markets in the Indo-Gangetic plains (Walton 1910, 44–45).

The collapse of agricultural commodity production and the dwindling trans-Himalayan trade weakened the regional economy of Uttarakhand. The only profitable economic activity that survived was the trade in timber and forest products, a sector that had not been directly promoted by the Company's policies. As the Company's reign in India came to be superseded by the Colonial Office in 1858, forest-based extraction had emerged as the dominant activity providing a growing share of income for many households in Garhwal and Kumaon. Some colonial officers were astute enough to see the discrepancies between the occupations reported to the census and the livelihoods actually pursued by households in the region. The district commissioner of Garhwal noted:

> The forests according to the census returns afford employment to 1,172, but this must be well below the real number. It is fairly obvious that, as practically every man in Garhwal is a *zamindar* [landholder], he has declared himself an agriculturist, ignoring his miscellaneous occupations which are often of much more importance. (Walton 1910, 64)

UTTARAKHAND UNDER COLONIAL RULE:
THE POLICY OF PROVISION, 1860–1914

The rise of timber extraction as a lucrative economic activity in Uttarakhand coincided with the growing concerns regarding the overexploitation of forests in India. State agencies that were involved in the development of infrastructure in the subcontinent, such as the public works department, worried that the rapid exploitation of forests would lead to depletion and increased expenditures for importing timber (Cleghorn et al. 1851). Revenue officers saw a potential source of state income that could be derived from taxing the unregulated forest extractive sector in a region that had seen a drastic fall in revenues (Ribbentrop 1900; Stebbing 1922; Baden-Powell 1892). Scientists and medical officers in the government's employ urged the state to restrict forest extraction to protect watersheds and guard against climate change (McClelland 1835). The pressure for government intervention in forestry steadily increased until 1864, when the Imperial Forest Service was created for the sole purpose of managing forests and regulating extraction.

The colonial government in India also faced a drastic reformulation of its dominant economic policy. As the Lancashire lobby grew more powerful in the British Parliament, and as Britain's financiers controlled the flow of money in the global economy of the nineteenth century, the colonial administration in India was compelled to adopt the policy of provision, which emphasized the steady flow of "essential supplies" for the industrial growth in Britain. Throughout the 1860s and 1870s, the colonial government redirected its policies to facilitate increased production and export of raw materials to Britain. Large regions were transformed into sites of cotton production; railways spread at a rapid pace, latticing the entire subcontinent and increasing the demand for timber used as sleepers (ties) for construction, and as charcoal for steam engines (J. Richards, Haynes, and Hagen 1985; R. Tucker 1983).

The competing demands made by different government agencies, along with those made by timber merchants, landlords, and village leaders, rendered the process of state forest management fairly difficult in Garhwal and Kumaon. Given the lack of alternative economic opportunities in the region, many households and small-scale traders had come to depend primarily on forest extraction for their livelihoods. Regulations that restricted the extraction of timber and other forest products were actively opposed by village leaders in the region. The colonial government sought to resolve these tensions by distributing the responsibility of forest management among different departments, leaving the Forest Service in charge of nearly two-thirds of the forested areas in the region.

The forests of Garhwal and Kumaon were reshaped according to new classifications: *reserved* forests, controlled by the Forest Service; *protected, civil, and unclassified wastes,* under the Revenue Department's authority;

village forests, controlled and managed by *panchayats;* and *private* forests, held by landlords and princes (G. Williams 1874; Ribbentrop 1900; Stebbing 1922). Yet despite this distribution of authority, forest classifications in Garhwal and Kumaon were vigorously contested. Between 1893 and 1914, forests in the region underwent three reclassifications. The key players involved in the disputes over forest classifications were the Forest Service, the Revenue Department, and local and regional elites who found their economic interests being eroded by forest laws and regulations (Brandis 1897; Rawat 1987; Stebbing 1922). The tensions between the two government agencies routinely arose over the issue of forest conservation and management. Forest Service officers complained that lack of conservation and management by the Revenue Department led to deforestation, soil erosion, and, in turn, increased pressures on adjacent *reserved* forests. The Revenue Department, which, unlike the Forest Service, allowed petty extraction and trade in timber and other forest products, received the active support of local and regional elites in opposing an extension of the Forest Service's authority over its forests (Rawat 1987; Ribbentrop 1900, 109–13).

By 1914, the forested landscape of Garhwal and Kumaon displayed remarkable differences in forest quality in the various classes of forests, mirroring to a large extent the social differentiation emerging in the region. *Reserved* forests, accessible to the few prosperous merchants in the region who had access to capital and credit for participating in large-scale forest extraction and trade, were relatively well stocked; the various subcategories of *civil* forests that were open to petty extractors and traders showed signs of rapid depletion; and those that remained accessible to a large proportion of poorer households for meeting their consumption and petty extractive needs were subjected to the highest pressures of resource extraction. The onset of World War I added to the prevailing demand for timber and increased the pressures on reserved forests in the region (Rawat 1987, 18; Stebbing 1932; for detailed analysis of the history of forest disputes in the region, see Rangan 1993a, 1995, 1997).

THE LAST PHASE OF COLONIAL RULE: RENEWED PROTECTIONISM AND INDUSTRIALIZATION, 1918–1947

World War I brought a whole set of complex problems facing the Indian economy to a head. The colonial government had struggled to manage a series of fiscal crises emerging from the costs of the Afghan War, its public works programs, famines, a global depression in the 1890s, and the depreciation of the Indian rupee, which had escalated the interest payments on its borrowings from financiers in Britain. All attempts made by the government of India to raise revenues by imposing tariffs on British imports were routinely quashed by the industrial lobby in Britain. In addition, India was expected to share the costs of World War I, which amounted to a "war gift" of Rs 100 million to the Home Government in Britain. In 1917, the government of India negotiated its contribution to the British treasury by procuring greater fiscal autonomy, and the freedom to increase tariffs on imports without the fear of retaliation (Dewey 1978, 42).[12] The war had revealed the strains that provisionist policies had imposed on the Indian economy. The growing military threats of competing European powers, along with the high costs of importing British goods, led the colonial government to reorient its policies toward industrial development in the subcontinent. Fiscal autonomy from Britain, demands for self-rule by the burgeoning nationalist movement, and the pressures from Indian industrialists gave added impetus to the government for embarking on industrialization policies.

The Forest Service in Garhwal and Kumaon enthusiastically responded to the colonial government's policy by setting up the Indian Turpentine and Resin Factory at Bareilly, a town bordering the Kumaon *terai* (the region between the Himalayan foothills and the Indo-Gangetic plain). Resin tapping had become an increasingly profitable activity when innovative methods for distilling resin and creosoting railway sleepers were developed around the turn of the century. The Forest Service succeeded in convincing the colonial government to transfer degraded forests under its control for *chir (Pinus longifolia)* plantations that would supply resin to the factory. It reassured laissez-faire adherents in the U.P. state legislature that private enterprise lacked the necessary capital for investing in a factory large enough to be profitable and prophesied a growth of employment opportunities in Garhwal and Kumaon. By the mid-1920s, the resin factory emerged as the largest supplier in India and began exporting to Europe (R. Tucker 1988, 98; Stebbing 1932, 594–99). Bobbin factories, sawmills, and turneries were established in the Himalayan foothills and towns in the terai. The newly established Forest Research Institute at Dehra Dun embarked on vigorous research of industrial uses of forest resources. Silvicultural practices and systematic forest botany developed by foresters during the nineteenth century in India were employed for experimenting with new combinations of traditional and exotic tree species in reserved forests. The institute became the leading forest research establishment and a model for similar centers that emerged in the British empire in Southeast Asia, Australasia, Africa, and South America (R. Tucker 1988, 99; Troup 1922).

The growth of the forestry sector and forest-based industrialization in Uttarakhand remained relatively modest during the interwar years. Timber prices collapsed in the late 1920s, and this was soon followed by the Great Depression of 1929. During the decade of the depression, strict fiscal discipline limited the budget for afforestation. This led the Forest Service to set aside vast tracts of reserved forests for natural regeneration and increased local animosity toward the agency. Protests erupted in various forms, ranging from litigation against the service to arson and poaching in reserved forests (Rawat 1987, 19–20; Stebbing 1932, 655–59). The growing power of nationalist leaders in local decision making reflected the changing balance of power in colonial India. Their influence was seen in the government's acceptance of the Kumaon Association's recommendations to the Forest Grievances Committee set up to inquire into the forest disputes in the region. When the financially strapped Forest Service called for labor contributions from local communities to assist in afforestation and routine conservation activities, village elites drew the support of nationalist leaders who protested against rendering unpaid labor services to the colonial government (Pant 1922).[13]

World War II increased the pressures of extraction on all classes of forests in Garhwal and Kumaon. The Forest Service increased extraction from reserved forests to meet the targets set by the War and Munitions Board (Champion and Osmaston 1962); routine management and conservation activities were set aside because of labor shortages. Local elites and petty commercial extractors made enormous profits from wartime scarcity given the relatively inelastic demand for timber, fuelwood, and charcoal (R. Tucker 1988, 102–4). By the end of the war and, soon after, the end of colonial rule in 1947, the Uttarakhand Himalaya bore scars of overworked forests. The region's landscape revealed both the high degree of dependence on the forestry sector as well as the inability or failure of the colonial government's policies to transform Uttarakhand's economy for the material well-being of the majority of its populace.

As the foregoing analysis shows, the ecological and economic transformation in the Uttarakhand Himalaya can better be understood by examining the processes of governance and institutional actions within dynamic spatial and political configurations. The focus on dominant policy phases reveals the multiple constraints and pressures faced by successive states as they altered their strategies to deal with changing global and regional economic conditions and shifts in the balance of class forces within their territorial communities. The ecological transformations in Uttarakhand between 1800 and 1947 emerged from necessary actions and contingent events faced by successive states, whose modes of governance had little to do with their apparently innate or immutable character. The policies and actions of pre-British Gurkha and Garhwali rulers were neither inherently superior nor inferior in character to those pursued by either the East India Company or the British colonial administration. The differences in their modes of governance and policy outcomes depended on the extent to which their actions were influenced by the interplay of political and economic processes beyond their immediate control, the material constraints and opportunities within their territories, and the extent to which social groups and classes within their jurisdiction redefined the exercise of control and governance.

NOTES

1 Shiva's writings seem to have a curious appeal to Western environmental audiences, not necessarily because her analysis displays theoretical rigor or careful attention to empirical detail, but rather because her portraits of Himalayan peasant women are remarkably essentialist. This essentialized representation strikes a chord with those who, it seems, are convinced that environmental salvation is only to be found in actions and practices of such women in the marginalized regions of the Third World.

2 For a detailed discussion of governance that differentiates between citizen and subject, see Mamdani 1996. Clearly, the definition of the citizen-subject-pastorate is itself a product of geographic history.

3 According to Traill (1828), nearly two-thirds of the trade revenue was from transit duties charged on the value of goods transported through their territories. The remaining amount was drawn from timber and mining royalties, taxes on grazing, and cesses on forest products, depending on the quality and extent of forested areas within the ruler's territory, and from taxing the services offered during the pilgrimage season.

4 Chetan Singh (1995) quotes Francisco Pelsaert, a Dutch traveler to India during the sixteenth

century who was struck by the high cost of firewood in the region, and the use of cow dung cakes as fuel substitutes by peasants.

5 The Company was granted the formal right and titular authority in 1757 to collect revenues and administer civil justice in Bengal by the Mughal emperor.

6 Wrigley (1978) notes that "though its name descends from the Middle Ages, the typical expression of the 'staple' policy was in the English Navigational Acts, and it constituted the 'mercantile system' which was denounced by Adam Smith in Book Four of the *Wealth of Nations*. Its beneficiaries are merchants, together with shipbuilders, dockers, hoteliers, and others whose living depends on an active foreign commerce" (23–24).

7 From the time of its earliest acquisition of territory in Bengal following the battle at Plassey in 1757, the Company's expansion proceeded steadily across northern India. Benares was acquired as an adjunct to Bengal in 1793, providing the Company the first foothold in the Indo-Gangetic plains. In 1801 the nawab of Oudh ceded areas north of Benares, the central Doab (*Doab* is a specific geographic term referring to the land lying between the Ganges and its major tributary, the Yamuna), and Rohilkhand; the Company conquered the upper Doab in 1803 and gained hegemony over large parts of the subcontinent by the time Garhwal was ceded in 1815 after the war with Nepal. Three years later, the Mahratta War was settled to the Company's advantage. In 1846, following the war with the Sikh rulers of Punjab, the whole coastline of India, excluding the Indus delta, was under Company control, with native states reduced to dependent powers. By 1856, a year before the Indian Mutiny, the Company had complete control over northern India, after having deposed Wajid Ali Shah, the nawab of Oudh. The entire territory that now comprises the state of Uttar Pradesh was called the United Provinces of Agra and Oudh.

8 These were Koth, Ráma Serai, and Gadoli in Garhwal, Lachmesar, Bhartpur, Hawalbágh, Rasiya, and Bhim Tál in Kumaun, and Kaulagir in Dehra Dun.

9 Six grantees joined to form a joint-stock agricultural company under the name of Maxwell, MacGregor and Co., consisting of forty shares with a paid-up capital of £20,000. The joint-stock company then proceeded to acquire additional land in the sub-Himalayan district of Sahéranpur, south of Dehra Dun across the Shiwaliks (G. R. C. Williams 1874, 314–32).

10 The Treaty of Lahore with the Sikh rulers of Punjab in 1846 launched a short period of peace that assisted the progress of canal construction.

11 The equivalent is £150,000, the conversion rate at the time being Rs 10 to £1. The estimate was based on charges levied at an average rate per acre of land irrigated, excluding revenues collected for mill rent, transit duties, and miscellaneous sales of canal produce (Cautley 1860, 28).

12 This victory coincided with the gradual weakening of the influence wielded by the Lancashire lobby and free-trade enthusiasts in the British Parliament.

13 Subsequent reports were published as "An Investigation into the Villagers' Rights in the Reserved Forests of Kumaon." See Rawat 1983, 1987.

J. Mark Baker

Colonial Influences on Property, Community, and Land Use in Kangra, Himachal Pradesh

British rule in the western Himalayan hill state of Kangra, which began in 1846, represented both continuities with, and disjunctures from, precolonial notions of sovereignty, property, and rule. Early colonial administrators, like their predecessors the Katoch rajas, were attuned to the importance of symbolic representations of state power. The early British revenue assessments in Kangra were also modeled after those of the prior Sikh government. However, the first land settlement of Kangra in 1850 facilitated changes in the control, use, and area of agricultural and forest lands in Kangra. These changes resulted from three interrelated processes. First, during the inherently contentious process of recording rights to land, settlement officers in this hill region applied models of property rights and the village "community" that had developed on the plains and that were informed by European notions of private property and agricultural development. The result was the creation of new "traditions" of land use and control. Second, Revenue Department officials emphasized the notion of property as a transferable economic resource that was allocated to individual property owners, in contrast to precolonial conceptions of property as an instrument for securing political legitimacy by distributing "interests" in property among different groups. Third, the Revenue and Forest Departments' use of land ownership as the sole criteria for assigning rights to forests and uncultivated areas increased local inequities; landless and nonagricultural groups were disenfranchised from resources to which they had previously possessed usufructuary rights of access and use.

Historical accounts indicate that precolonial property rights in hill areas significantly differed from those on the plains, especially in terms of the absence of a coparcenary village "community" and locally controlled and managed common lands *(shamilat)* (Baden-Powell 1892; Barnes 1855; Douie [1899] 1985). Notwithstanding these differences, Revenue Department officials during the first settlement applied models of land tenure and the "village community" from the plains to the hill areas. In addition to granting alienable rights to cultivated areas, the settlement created village common property (shamilat) where previously it had not existed and then allocated private rights to shamilat resources based on land ownership, in the process annulling the usufruct rights of landless households. Revenue Department officials, in the interests of promoting agricultural expansion, also granted extensive rights in the soil and forests of uncultivated areas to revenue-paying landowners. However, because of shifting policy objectives soon after the first settlement, the Revenue Department—and later and more aggressively the Forest Department—sought to reestablish territorial state authority and control over as extensive an area as possible through forest demarcation and the restriction of local rights to forest resources. In contrast to the competition between local and state forest use that developed during the late nineteenth century, the colonial administration played a facilitative role in promoting agricultural intensification. To this end, the civil administration subsidized the construction and repair of some of the largest gravity flow irrigation systems, known as *kuhls,* in Kangra and occasionally adjudicated water conflicts. The forms of property rights the British developed in Kangra reflect the confluence of utilitarian notions of agricultural development and private property, the experiences of Revenue Department officials in the plains, and a selective reading of local tradition and custom, interacting with shifting colonial priorities that first emphasized agricultural expansion and later forest conservation.

PRECOLONIAL PROPERTY STRUCTURES
AND PROPERTY RELATIONS IN KANGRA

The political organization of Kangra hill state in the late eighteenth century conforms to Cohn's (1987) model of a little kingdom that at various times and to various degrees was subject to the rule of Mughal successor states. Because of Kangra's unique position close to the routes linking Central with South Asia, the degree of autonomy of its ruling Rajput lineage was determined by who controlled the plains to the south and west and by the strength of their control. When regional political regimes—whether centered in Delhi, Lahore, or Kabul—were weak, the Katoch chiefs ruled as independent sovereigns in Kangra and at times reduced neighboring hill states to the status of their tributaries. When a plains-based central government consolidated its authority and rule and extended it over the Rajput-ruled hill states in the western Himalaya, the Katoch rajas themselves became tributaries.

Throughout the precolonial period, the control and exchange of land were used to consolidate, strengthen, and maintain political power. For a ruler, whether a Katoch raja, Sikh chieftain, Mughal emperor, or Afghan *durani,* the importance and meaning of land derived from its use as a medium of exchange for negotiating sovereign claims to territory and for transforming potential adversaries into political allies.[1] Rulers made gifts of land to individuals who served the ruler in a military, administrative, or other capacity (Habib 1963; Lyall 1874). Rulers gave land to religious institutions such as temples or mosques and to religious leaders, theologians, and saints (Dirks 1992; Habib 1963). Land grants were also given to potentially rival political groups to secure their support.

When a ruler assigned a grant of land to an individual, temple, or shrine, he transferred the right to the revenue from that area; the existing array of occupancy, cultivation, and other usufruct rights was left untouched. This conforms to the precolonial model of property relations based on overlapping sets of interests rather than on exclusive claims of ownership (Embree 1969). Habib suggests that in such areas a "single owner cannot be located" and that instead one finds the allocation of "different rights over the land and its produce, and not one exclusive right of property" (1963, 118).[2]

However, as Dirks (1992, 179) points out, land was only one of many mediums of exchange by which a ruler secured political legitimacy and maintained sovereignty. Dirks suggests that "the king ruled by making gifts, not by administering a land system in which land derived its chief value from the revenue he could systematically extract from it." In addition to gifts of land, rulers in Kangra distributed the hereditary right to cultivate a particular field, the rights held by transhumant Gaddis and Gujars to particular grazing runs, the right to some inherited administrative posts, and the right to operate a water mill or to erect a fish weir. These rights were all called *warisis* and derived from the raja as a separate, taxable tenancy (Barnes 1855, 18–19; Lyall 1874, 17, 4–5; Baden-Powell 1892, 693–94). The proof of entitlement to a particular warisi was the *patta,* a deed that spelled out the rights and responsibilities that the warisi entailed and the terms by which it could be renewed. Maintaining monopolistic control over the power to grant a patta was central to the ruler's sovereignty. Lyall (1874, 21) mentions that the rajas "jealously" guarded their monopoly to grant pattas for warisis—"under them [the rajas] no *wazir* or *kardar* [administrative officers] could give a *pattah* of his own authority." To do so could threaten the legitimacy of the ruler.

Unlike tenure systems in the plains—where village boundaries invariably included uncultivated areas of common land used for grazing and the locally regulated rights to those resources were a function of residency and land ownership (Kessinger 1974)—use of the uncultivated lands in Kangra appears to have been more informally regulated and at least nominally at the ruler's discretion. All individuals, regardless of whether they held rights to cultivated land or were even agriculturists, possessed usufructuary rights to uncultivated areas for subsistence purposes (Douie [1899] 1985, 69). These rights included the right to graze livestock, cut grass and leaves for fodder, remove thorns for hedges, and collect dry wood for fuel (Lyall 1874, 19). These usufruct rights were generally subordinate to the right of the ruler to grant a patta to an individual to bring a section of the uncultivated area into cultivation. However, Singh (1998, 95) does note that farmers could object to such a grant if it was made to an individual from another village, especially if they had already requested to extend cultivation to that area.

Lyall, in his revision of the original land settlement, and perhaps seeking legitimacy in history for asserting increased colonial rights in forests in Kangra, notes that precolonial rulers claimed all rights to forests and that they recognized two classes of forests. In forests reserved as shooting preserves, no grazing or collecting fodder was allowed. In other forests, individuals could cut timber for roofing or the construction of an agricultural implement, and for marriage and funeral ceremonies (Lyall 1874, 19). In forests where grazing was permitted, the ruler sometimes imposed a ban on grazing during the monsoon (Lyall 1874, 21). Lyall postulates that in addition to benefiting "trees and game," the imposition of a grazing ban served as an assertion of the ruler's authority and, for Lyall's purposes, provided a legitimizing model of strong state control of forest resources.

The precolonial Katoch rulers also sponsored the construction of nineteen of the longest and most complex gravity flow irrigation systems *(kuhls)* in Kangra. Most of these kuhls were named after the raja or rani who sponsored them. At least seven of these kuhls were constructed during the late eighteenth century at the apogee of Katoch rule (*Riwaj-i-Abpashi* 1918).

State sponsorship of kuhls was a means, much like land grants, whereby a ruler could strengthen sovereign control over a region. In exchange for sponsoring the construction of a kuhl, the ruler augmented the amount of grain the state received in assessed rent and increased his political legitimacy and authority. This legitimacy was strengthened because not only cultivators but artisans, traders, and others benefited from the construction of a kuhl, for it satisfied household water needs as well as the water requirements for livestock and the small kitchen garden found in most domestic compounds. The kuhls rulers constructed were also named after them, thereby ensuring that they would be remembered long after their death. In some cases, kuhl sponsorship was linked with religious patronage, thus strengthening the divine mandate of the ruler.

THE PRECOLONIAL "VILLAGE COMMUNITY"

In Kangra, rights in cultivated land were not part of a joint estate shared in common by a kin group (clan or lineage), as was often the case on the plains. Instead the hereditary right to cultivate land was "owned" by families and derived from the patta, received from the ruler. Although hereditary forms of proprietary rights were not easily transferred, they did constitute a strong and defensible claim to cultivate an area (Singh 1998, 57). Revenue was assessed on the basis of cultivated area per family holding, not at the village level. The landholders within a village had no corporate responsibility for revenue payment. If the rent from a family holding was not paid, the ruler's agent took action against the individual holder only; "the other landholders of the circuit had nothing to do with the matter; each plot or holding was a little *mahal* [place] of itself" (Lyall 1874, 28).

The lack of locally appointed village headmen and village *panchayats* further underscores the absence of coparcenary shareholders. In most plains villages, the village headmen were local men whose primary responsibilities and authorities included revenue collection, maintenance of civil order, and the allotment of uncultivated land to prospective cultivators (Habib 1963; Kessinger 1974). In Kangra, headmen were appointed by the ruler or his agent, not by the inhabitants of a village. The headman's primary responsibility was the collection of land revenue. He did not have the authority to allocate the right to cultivate uncultivated land to other individuals.

The absence in the hills of coparcenary village communities was described by the early British settlement officers. For example, Baden-Powell notes that "in the hills . . . villages, in the proper sense of the term, hardly exist; we have merely aggregates of a few separate holdings . . . in reality the management is as nearly raiyatwari [peasant cultivated and owned] as possible," and "we find no 'villages' in the ordinary Indian sense, and consequently . . . no joint-proprietary communities over villages" (1892, 2:537, 692). Barnes, referring specifically to Kangra, wrote that

[the inhabitants of the hill village] have no community of origin, but belong to different castes. There is no assemblage of houses like an ordinary village, but the dwellings of the people are scattered promiscuously over the whole surface. Each member lives upon his own holding, and is quite independent of his neighbor. There is no identity of feeling, no idea of acting in concert. The headman, who is placed over them, is not of their choice, but has been appointed by the Government. In short, the land enclosed by the circuit, instead of being a coparcenary estate, reclaimed, divided and enjoyed by an united brotherhood, is an aggregation of isolated freeholdings quite distinct from each other, and possessing nothing in common except that for fiscal convenience they have been massed together under one jurisdiction. (1855, 16)

However, the effects of "massing" holdings together for the sake of convenience during the first settlement had considerable effects on the control and use of uncultivated areas.

THE FIRST BRITISH SETTLEMENT OF KANGRA

The early years of British rule in Kangra were marked by strong continuities with, and disjunctures from, precolonial administration. The first summary land settlement made in 1846 immediately following the forcible annexation of Kangra from the Sikhs was only a slightly modified version of the Sikh government's prior system of revenue collection. And in some important respects, the first regular settlement by Barnes in 1850 was also modeled after the Sikh system. Symbolic continuities with prior ruling regimes also existed. For example, British officers chose Kangra Fort as the first district administrative headquarters. Barnes writes that although the fort had a garrison and was close to the town of Kangra, the main reason for choosing the fort was "the prestige attaching to the name . . . the same spot which had ruled so long the destinies of the hills still continued to remain the seat of local power,—the center whence order emanated, and where supplicants repaired for redress" (1855, 15).

Despite drawing on local referents that emphasized continuity with previous state regimes and helped to legitimize their own rule, the British were self-consciously aware of the marked difference of their rule. Barnes writes at the conclusion of his account of the history of Kangra state, "I turn with pleasure from the narrative of wars and insurrections to the quiet details of our administration and the general statistics of the district" (1855, 15). The pen of British authority was extensively used in the first regular settlement to codify previously unwritten customs and rules and to confer new rights. Within the space of only a few years, "the quiet details of . . . administration" wrought significant changes in property rights in Kangra.

The Influence of Settlement on Land Tenure and Agriculture
After four years as district commissioner, G. C. Barnes conducted the first regular settlement. In determining the rates of assessment, Barnes was guided by the rent rolls from the prior Sikh government and the summary British settlement. In fact, as Lyall pointed out twenty years later, Barnes's settlement was "nothing more than the old native assessment very slightly modified" (quoted in A. Anderson 1897, 12). The primary changes consisted of a reduction in assessment rates on unirrigated areas and the removal of the host of extra taxes that had accompanied the land revenue.

However, although the rates of taxation between the precolonial and colonial regimes were remarkably consistent, the methods of tax payment and the nature of property rights were significantly altered. Barnes, using the system prevalent on the plains, shifted responsibility for paying land revenue from individual families to all of the landholders in a revenue village, who were then made jointly responsible for the revenue. As Lyall observed twenty years later, this "bound together the landholders of each *mauzah* [revenue village] into a kind of village community" (1874, 28). Barnes also reversed the "ancient and time-honored custom" of paying rent in kind by commuting in-kind to cash payments (1855, 52). The switch from in-kind to cash payments was part of the then prevailing utilitarian philosophy of agricultural development in Europe. That Barnes embraced this philosophy is strongly suggested by his comments on the effects on farmers of substituting cash for in-kind payments: "It has taught them habits of self-management and economy, and has converted them from ignorant serfs of the soil into an intelligent and thrifty peasantry" (1855, 52).

Although he did not explicitly acknowledge this, by conveying the same rights in land that had been granted to landholders in the North-Western Provinces, Barnes conferred full proprietary rights in cultivated land to individuals who had previously held an inheritable but not alienable right. Subsequent Revenue Department officers have argued that the introduction of the right to alienate land was an unintended consequence, a "mere incident of the (first) settlement" (A. Anderson 1897, 9). Whether intended or not, this new right had long-reaching consequences once rights holders realized they had been granted the power of alienation. By 1890 14 percent of the total cultivated area was under mortgage, and an additional 5 percent had been sold (A. Anderson 1897, 9). Settlement officers unanimously attributed the increase in alienated land to the need to raise capital for bride-price payments among high caste Rajputs as well as upwardly mobile Rathis and Thakurs seeking to legitimize their claims to Rajput status (the dowry system only relatively recently supplanted the prior bride-price arrangement) (A. Anderson 1897; Connolly 1911; Middleton 1919). During this period, the price of a bride increased tenfold from Rs 20–40 to Rs 200–400 (Parry 1979, 243).

The increase in land transactions during the latter part of the nineteenth century was made possible by the recent changes in property rights and by the introduction of a bureaucratic set of procedures and laws (e.g., district courts, land laws, codified statements of rights regarding land ownership, etc.) that provided the means for transferring property from one owner to another.[3] Urban-based moneylenders and other urban groups began to accumulate land under these laws. The Punjab Land Alienation Act of 1900 was an attempt to slow these forms of land transfers by prohibiting "nonagricultural" castes from purchasing agricultural land. In Tehsils Dehra and Hamirpur of District Kangra, for the nine-year period preceding and following passage of the land act, the percentage of total cultivated land mortgaged declined from 13.8 to 3.8 and from 9.5 to 3.5 percent respectively (Connolly 1911, 6).

Following the first settlement, the cultivated area in Kangra also increased. By 1890 it had increased 8 to 10 percent. Hill slopes that had been infrequently cultivated previously were terraced and cultivated annually, and some forested areas were converted to agriculture. This agricultural expansion was facilitated by shifting the authority to control the expansion of agriculture from the ruler to the landholders of a hamlet (accomplished as part of the first regular settlement), offering the financial incentive for intensifying agriculture provided by twenty-year fixed assessments, and increasing grain prices.

Expanding irrigation also accompanied the increase in cultivated area. In Palam subdistrict, between 1851 and 1890, 146 acres of uncultivated area were converted to agriculture and irrigated by constructing new kuhls and extending preexisting kuhls (O'Brien 1890, 14). Between 1850 and 1916, forty-one new kuhls were constructed in Kangra Valley (*Riwaj-i-Abpashi* 1916). In 1855 Barnes observed that after the first settlement, single-cropped fields were double cropped, and kuhls were "projected and executed" (1855, 63). In 1897 Alexander Anderson remarked on the "new watercourses" that had been constructed since Lyall's first revision of the settlement in 1874 (A. Anderson 1897, 60). And Middleton, in the introduction to the *Riwaj-i-Abpashi,* noted that the records pertaining to kuhl irrigation drawn up during Lyall's revised settlement were no longer accurate owing to post-1874 judicial verdicts and to new kuhl construction (*Riwaj-i-Abpashi* 1916).

The right to alienate land granted to landowners in the first regular settlement of 1851, coupled with subdivision through inheritance, resulted in some tenant cultivators becoming owner cultivators and decreased the overall average landholding size. In Palam subdistrict, of the 2,060 hectares sold by landowners between 1871 and 1890, 14 percent was sold to individuals who were landowners in 1871 and who were also moneylenders, 41 percent was sold to individuals who were landowners in 1871 and who were not moneylenders, 20 percent was sold to new agriculturists who did not own land in 1871, and 25 percent was sold to European tea planters (O'Brien 1890, 4).[4] In Tehsil Palampur, between 1871 and 1890, the num-

Table 1: Percentages of Cultivated Land, District Kangra, by Caste, by Tehsil, 1919

Tehsil	Brahmins	Rajputs	Rathis	Girths	Others
Palampur	21.0	20.6	17.5	9.1	31.8
Kangra	11.5	21.3	5.8	33.1	28.3
Nurpur	12.6	55.1	10.0	2.5	19.8
Total for District Kangra	14.9	35.6	11.3	12.4	25.8

Source: Middleton 1919, 3.

ber of small landowners increased from 13,854 to 22,081, the number of medium landowners declined from 5,553 to 5,178, and the number of large landowners declined from 1,064 to 716.[5] During this same period, the average size per holding for small landowners decreased from 0.8 to 0.6 hectares, for medium landowners it remained constant at 3.0 hectares, and for large landowners it increased from 9 to 10.4 hectares (O'Brien 1889, 1890, 1891a, 1891b). Many of the large and medium holdings belonged to European and local tea planters (O'Brien 1890).

In the decades following the first settlement, absentee landownership increased, and consequently so did the tenant-cultivated area. In 1874 tenants cultivated 92,003 hectares (19 percent) of the cultivated area in District Kangra (Lyall 1874, 86–93).[6] By 1897 the tenant-cultivated area in the district had increased by 71,000 hectares to 163,203 hectares, or 32 percent of the total cultivated area (A. Anderson 1897, 5). During this period, the total cultivated area of the district increased from 480,704 to 509,579 hectares, and the owner-cultivated area decreased from 388,739 to 344,803 hectares. The decrease in owner-cultivated area, when summed with the increase in the district's cultivated area, is roughly equivalent to the increase in tenant-cultivated area. These trends suggest that the increasing numbers of small landholders were also tenants because their average landholding size was too small to maintain a household. This could account for the increasing tenant-cultivated area between 1871 and 1897. The decreasing owner-cultivated area suggests land transfers from owner cultivators to members of moneylending castes such as the Mahajan as well as increased out-migration for employment.

The primary landowning castes during this period were (and continue

Table 2: Land Ownership, by Caste, Kangra Valley, 1890

	TOTAL FOR KANGRA VALLEY			
	No. Holdings	Cult. Area (ha)	Avg. Holding (ha)	% Total Holding
Girths	5,466	9,676	1.8	14.4
Brahmins	5,869	16,723	2.9	25.0
Mahajans	1,282	4,194	3.3	6.3
Rajputs	5,557	15,259	2.8	22.8
Rathis	4,061	11,827	2.9	17.6
Others	6,332	9,344	1.5	13.9

Source: O'Brien 1889, 1890, 1891a, 1891b

to be) Rajputs, Brahmins, Rathis, Thakurs, and Girths. Table 1 shows the ownership of cultivated land by caste for District Kangra in 1919, and table 2 gives land ownership statistics by caste for Kangra Valley in 1897. Table 1 indicates that together, the two highest castes (Brahmins and Rajputs) owned just over 50 percent of the total cultivated land. In 1931 they made up 44.2 percent of the district's population. Table 2 demonstrates a similar trend in Kangra Valley and also illustrates the relatively small differences in average landholding size between castes.[7]

The Influence of Settlement on Property in Kuhls

Consistent with their interest in promoting agricultural expansion, the Revenue Department also facilitated, subsidized, and generally supported the expansion of irrigation networks. This process involved regulating the construction of new kuhls, codifying irrigation customs and rights, mapping kuhl networks, and shifting dispute resolution from the village level to the district courts.

Settlement officers asserted that state claims to the natural waterways of the district represented a continuity with, rather than a change from, previous customs. Lyall (1874, 56) wrote:

In order to retain in its hands the power of making new irrigation channels where needed, the Government directed all Settlement Officers to assert its title to all natural streams and rivers. In Kangra the title of Government, by old custom of the country, was particularly clear, and I accordingly asserted it.

Permission to construct a new kuhl could not be granted unless the government had a record of the existing network of kuhls. Irrigation rights were first recorded in the second settlement. Maps were drawn of every stream showing the position of each kuhl, its headworks and main channels, and the villages through which it flowed. Appended were attested records of the customary rules regulating the relations between communities that share one kuhl regarding water distribution, the manner of constructing the headworks, responsibility for repairs and maintenance, and a short history of the kuhl. A glossary of specialized irrigation terms was also included. These statements of rights were bound, and copies were kept at the Palampur and Kangra Tehsil offices. They constituted the first edition of the *Riwaj-i-Abpashi* (Irrigation Customs).

The *Riwaj-i-Abpashi* represented the first time that complex irrigation customs guiding the measurement and distribution of a single kuhl's water to as many as sixty different hamlets were reduced to writing. Settlement officers determined the irrigation customs and practices relating to a specific kuhl by calling a public meeting and asking those present to describe their customs and practices. After writing them down, the officers read them aloud and incorporated suggested changes, and then prominent village leaders attested to the veracity of the statement with their thumbprint or signature. The resulting document constituted a legal record of rights that was, and still is, used as the basis for judicial decisions regarding water disputes.

Lyall acknowledges the difficulty of creating an accurate statement of irrigation rights in this manner. He notes that "probably these statements are sometimes incorrect. . . . the custom is often vague and difficult to define" (1874, 243). If irrigation customs appeared vague to a settlement officer, one wonders if they appeared equally vague to the shareholders whose irrigation water depended on them. Or factions well represented at the general meeting may have presented the settlement officer with a picture of rights in a kuhl that favored their own interests. The complex effects of codification aside, the creation of the *Riwaj-i-Abpashi* did constitute a new template against which the civil administration would resolve subsequent conflicts over irrigation rights and responsibilities.

The civil administration also indirectly supported kuhl regimes by helping to reconstruct kuhls following natural disasters, and by adjudicating water conflicts during periods of water scarcity. On 4 April 1905 Kangra was devastated by a severe earthquake that wreaked havoc throughout the district; 12,663 people died in Kangra and Palam Tehsils alone (Middleton 1919, 5). In addition to leveling most structures in the valley and destroying roads and bridges, the earthquake occurred at the beginning of the rice planting season, when labor demands and dependence on kuhl water were greatest. To avert a localized famine, the administration mobilized the military to restore transportation links and rebuild the destroyed kuhls.

The colonial government also sponsored the construction of two new kuhls. One was built to supply water to the growing town of Palampur. The British were interested in developing Palampur as a trading center with Afghanistan and eastern Turkestan, where they sought a market for Kangra tea. The second kuhl was constructed to provide water to the town of Dharmsala, a hill station and headquarters of the district administration.

The Influence of the Settlement on Property in Uncultivated Areas
Barnes, in the first regular settlement, implied—and subsequent government decisions affirmed—that landholders had been made coproprietors of the uncultivated areas to which they had previously held only usufruct rights. However, Barnes did not dwell on this point because, as Lyall speculated twenty years later, he did not want to draw attention to the fact that he had "effected a revolution in the old state of property," something a settlement officer was not empowered to do (Lyall 1874, 19). Barnes converted the landholders of each hamlet into a coproprietary class and transferred to them ownership rights in the uncultivated areas, known as the waste. This transfer of property had many implications. It nullified the rights of landless households to forest resources collected from unenclosed uncultivated areas. Revenue from these areas, previously paid to the ruler, was now collected by the *lambedar* (village tax collector) and distributed to all landholders in proportion to the amount of revenue each paid. And now landholders, rather than the state, had the authority to grant permission to an individual to reclaim and cultivate an uncultivated tract.[8]

The transfer to landholders of rights in these areas may have been an unintended consequence of the application of land use categories from the plains to the hill states, or it may have been an intentional, if implicit, policy to promote agricultural expansion and intensification by simplifying the process of bringing new areas under cultivation and more intensively cultivating already cultivated areas. The statement of rights created during the settlement operations for each hamlet was the first instance of previously orally transmitted customary rights and practices being set to writing. It constituted a new arena in which social groups could assert competing claims to resource access, control, and use, and it provided opportunities for Revenue Department officials to import new land use categories and forms of property. For example, Barnes noted that in the preparation of the record of rights for each hamlet, he gave the subject headings and elicited information with questions and even suggestions. Furthermore, he instructed the *tehsildar* "to write down the actual practices as observed . . . and not to fill up details after his own imagination" (1855, 67).

The history of the term *shamilat,* referring to village common property, in Kangra exemplifies this process. Shamilat was first introduced into the district as a land use category during the first regular settlement. Shamilat was imported from the plains; it had no pre-British referents in Kangra. Twenty years after the first settlement, Lyall argued that landholders had not manufactured their own title to the wastes by putting shamilat in the record of rights, but rather that "the real inventors of the definition [of shamilat] were the native officials and clerks who worked under Mr. Barnes" (1874, 31) who had inserted "Shamilat" as the heading in the village records.[9] Whether the creation of Shamilat as a land use category was intentional or not, it did encourage the expansion of agriculture by granting landholders the right to break up and cultivate the waste, free of extra revenue, for the duration of the settlement.[10]

Following the transfer of rights in uncultivated areas from the state to coproprietary landholders, Revenue Department officials attempted to privatize as much of it as they could. In the following passage, Barnes describes how he approached areas declared as shamilat:

> Whenever . . . I saw an opportunity, I insisted on a partition of the estate according to the number of shares. Every inch of profitable ground was divided and allotted to one or other of the co-partners. I ignored as far as my means would allow the very name of "*Shamilat,*" for experience has assured me that the smallest portion left in common will act as a firebrand in the village. It is sure to lead to dis-

sension, and forms, as it were, a rallying-point for the discontented and litigious to gather round. (1855, 67)

Similarly, during the second settlement, Lyall allocated as much of the waste as he could to individual households. Uncultivated areas such as hay fields, hedgerows, and plots surrounded by individually held property, when found "in the exclusive occupation and possession" of individuals, were also recorded, with the consent of the village, as the private property of those individuals (1874, 218).

The Influence of the Settlement on Property in Forests
Early colonial forest policies in Kangra reflected a mix of shifting policy objectives regarding the distribution of rights and obligations between state and society, as well as growing tensions between the Revenue and Forest Departments regarding the control of Kangra's forests. At the time of the first settlement, the primary policy objective, secondary to forest conservation, was to encourage agricultural expansion into uncultivated areas (Tucker 1982, 115). To facilitate agricultural expansion, the first settlement emphasized local control over the district's forests. Rights to forests were "partially assigned" to the landowners of a village to provide local incentives for forest protection and to pay the forest watchman (Lyall 1874, 30). In addition to providing local incentives for forest conservation, this approach eliminated the need to develop the institutional capacity for forest conservancy and saved the colonial administration the expense of surveying the forests and identifying the various right holders.

However, less than five years after the first settlement, demand for land for tea plantations by British planters (Rangan, this volume) and the increasing value of timber for railroad sleepers (ties) led to a shift in forest policy in Kangra. District civil administrators attempted to reinterpret the intent of the initial settlement report in such a way as to recoup government rights to forests and uncultivated areas that had been "lost" during the first settlement, when such rights had not been valued. They did this to claim the right to allocate uncultivated areas to British tea planters and to assert government ownership and control over increasingly valuable forest tree species. Although the chief commissioner of Punjab rejected these early attempts on the part of Kangra district officials to "reclaim" lost forest rights, subsequent district Revenue Department officers were able to implement a series of increasingly restrictive regulations governing access to forest resources. These culminated in the demarcation of forests in Kangra in a process that Vandergeest and Peluso (1995) have called "the territorialization of state power." By restricting what actions could or could not occur within a bounded and mapped spatial area, the colonial government explicitly linked territorial control with state authority. The second district commissioner of Kangra began this process by introducing new rules governing forest access and use in Nurpur and Dehra subdistricts. The new rules divided the forests within village boundaries into three parts. In one part the Forest Department annulled all customary rights to collect forest products and kept it as a reserve for a rotational period of at least three years. In the remaining two parts, the Forest Department also curtailed forest use by making it a penal crime to burn grass in the winter to improve subsequent fodder production. The Forest Department also required, contrary to the first settlement, that a landowner seek permission from the deputy commissioner to break up and cultivate any unenclosed uncultivated area.

The whittling of local rights in forests continued. In 1859 landowners who previously sought permission from the hamlet tax collector to cut green timber for building or agricultural purposes were now required to request permission from the subdistrict officer. They also had to pay for the timber, whereas previously they had received it gratis. In exchange, 25 percent of the value of the timber sold by the government was returned to the village from which the timber was cut. Three-sixteenths of the total sale went to village officials, one-sixteenth to the community of landholders, and none to landless households. Similarly, the district commissioner reserved the right to refuse permission to an individual to break up and cultivate uncultivated forest areas, even if the applicant offered to pay the value of the standing timber. This permission was generally refused, except when requested by Europeans who wanted to plant tea gardens or cinchona groves, for which cases the government "saw good reason for sacrificing its forest rights" (Lyall 1874, 238).

The Indian Forest Service, created in 1865, generally competed with the Revenue Department for control of forest lands (Rangan and Saberwal, this volume). The Revenue and Forest Departments also tended to manage forests for different packages of forest uses, with the former emphasizing the provision of forest products to meet local subsistence needs and the latter emphasizing the production of commercial forest products, especially timber and resin. In 1869 the Forest Department succeeded in acquiring from the Revenue Department the forests of Nurpur, Dehra, and Hamirpur Tehsils. However, the Revenue Department retained control over the forests in Tehsil Kangra because most of the forest area had been purchased by tea planters or was not available for forest conservancy because of proximity to towns and stations. Acting on the suggestions of Baden-Powell, the Forest Department in Nurpur and Dehra Tehsils attempted to demarcate and acquire absolute rights to some forests in exchange for permanently relinquishing some rights in the remaining forests to village communities. The demarcated forests obtained in this manner were declared reserve forests in 1879.

This method of acquiring absolute rights to small, isolated patches of forest was considered unacceptable to Forest Department officials and hence not applied beyond Nurpur and Dehra Tehsils. In the remaining *tehsils* of the district, the conservator of forests sought to achieve control over a greater area of forests than had been achieved in Nurpur and Dehra. To this end, the conservator called for a general demarcation of the forests in Kangra according to the following principles: (1) there would be no "give-and-take" negotiations with landholders over rights of access and use, (2) no change was to be made in forest management either inside or out of the demarcated areas, and (3) forest areas that might be broken up for agriculture should be demarcated separately from forest areas that would be permanently maintained as forests (A. Anderson 1887, 2). By 1885 the demarcation of all the forests in the district as either reserved or protected was completed. A total of 257.5 square miles of forest was demarcated, of which 75.5 square miles were closed against grazing and other usufructuary rights (A. Anderson 1887, 3).

By demarcating areas as reserved forest, the Forest Department ensured that those areas would not be converted to agriculture in the future. The process of demarcation gave the Forest Department an opportunity to further restrict the access to, and use of, forest resources within demarcated areas. The department made the payment of land revenue the only basis on which valid claims to forest resources could be established in the record of rights. This nullified the claims to forest resources of all landless individuals and families who had previously possessed customary use rights to the forests within the demarcated areas.[11] Landless individuals without rights to forest resources collected and sold wood, charcoal, and grass in the small towns of the district. The demarcation and associated rules also sought to control this illegal market in wood and grass by prohibiting it or requiring a license to sell these products.

The Forest Department also restricted the cutting of green brush and the tree species that could be cut for building purposes. Cutting green brush for noncommercial purposes had previously been allowed after asking permission from the village tax collector. In practice, however, this rule was rarely enforced, and in 1864 the deputy commissioner, whose authority superseded that of the district Forest Department official (Saberwal, this volume), passed an order that there would be no restriction whatsoever on the cutting of brushwood. Had this order been implemented, it would have undermined Forest Department authority, but as Anderson happily notes, in the forest settlement, "fortunately the order was not made generally known, and certainly was not acted upon" (1887, 8). During demarcation the Forest Department reintroduced the restriction on green brushwood cutting, admittedly more as a demonstration of its territorial authority than for erosion control. However, even the Forest Department was aware of the limits to which they could go before facing social unrest. Anderson suggests that whereas requiring the permission of the subdistrict or forest officer would be preferable, doing so would "raise a storm of opposition" and therefore was not advisable. Similar restrictions were imposed governing which tree species could be cut for building or agricultural purposes. A list of sixty-two tree species was drawn up for which the right to cut would only be given on payment to either the tax collector or the forest officer. Local opposition to this rule was strong enough to compel officers of the Revenue Department to ask forest officers to reduce the number of restricted tree species. Other examples of the restricting impacts of demarcation include the instigation of charges for trees used in wedding and funeral ceremonies, limits on the number of trees that could be claimed for building purposes per year, and the types of activities considered legitimate building purposes.

CONCLUSION

The property rights regimes instituted by colonial administrators represented attempts to define a colonial vision of the state and its relations with local communities, an effort that simultaneously helped create a specific type of local "community." The property rights in cultivated, uncultivated, and forested areas the colonial administration promulgated and the commutation of in-kind to cash payments were informed by a combination of contemporary utilitarian ideas of agricultural and economic development, prior experience with the Northwest Frontier Provinces settlement, shifting policy priorities, and conflicting mandates and power struggles between the Revenue and Forest Departments. The first settlement granted cultivators the right to alienate the land they tilled, the right to cultivate previously uncultivated areas, and substantial rights of forest access and use and made a village "community" collectively responsible for revenue payments. These changes, combined with the initial policy preference for agricultural expansion over forest conservancy, a fixed twenty-year settlement, and rising grain and bride prices, resulted in unprecedented land transfers and mortgages, the expansion of cultivated areas at the expense of forestlands, the emergence of a noncultivating mercantile elite, an increase in tenant-cultivated agricultural lands and absentee landownership, and an increase in the number of small landholdings.

Concomitant with the expansion of agriculture during this period was the increase in the area irrigated by kuhls. As had the precolonial regime, so the colonial civil administration facilitated the expansion of irrigation networks. Although the facilitative role of the colonial administration in water management represented more a continuity with precolonial rule than a disjuncture, the role of the colonial state in forest management differed sharply from that of its precolonial predecessor and from its own role in irrigation management. Soon after the first settlement, forest conservancy began to compete with agricultural expansion as an administrative policy goal. To secure more complete control over forest access and use, a series of regulations and orders was passed, first by the Revenue Department and later by the Forest Department. These forest rules whittled away at local usufructuary forest rights and eliminated those of nonlandowning households, thus underscoring the fact that forest demarcation can also be a form of community demarcation.

This chapter has benefited from the constructive comments of the other contributors to this volume, the volume's commentators, Kim Berry, Ruhi Grover, and two anonymous reviewers. Financial support for research was provided by a Ciriacy-Wantrup postdoctoral fellowship at the University of California at Berkeley.

1 See Neale 1969 for the meaning of land control in precolonial India as a source and instrument of political power, in contrast to British conceptualizations of land as an input within an economic system whose internal logic is profit maximization. Further analyses of the political and social functions of land in precolonial India are found in Embree 1969; Dirks 1985, 1992; and Cohn 1987.

2 For example, interests in (i.e., claims to) the produce of a field are distributed among the cultivator; the various village artisans such as the potter, basket maker, iron smith, and water master who receive a portion of the harvest as compensation for the services they provide; the individual who may hold the hereditary right to cultivate the field but engaged with another for the fields' actual cultivation; and the individual who claims the right to the revenue from the field.

3 Under certain conditions, land transfers also occurred during the precolonial context. Land could be transferred by gift if the patta holder had no heirs. Similarly, proprietors in arrears of revenue could mortgage their land to another individual, who would then be responsible for paying the revenue and in exchange would receive half of the harvest from the former proprietor-cultivator. In some cases, if the arrangement became long term, or "by error at [the] first settlement," the former proprietor's claim to the land was reduced to that of a tenant, and the mortgage holder became the proprietor (Lyall 1874, 66). Although these forms of land transfers did occur, the hereditary right to cultivate land was not bought and sold as a commodity before the first regular settlement (Barnes 1855).

4 The 2,060 hectares sold between 1871 and 1890 represent 12 percent of the total cultivated area in Palam Ilaqa (O'Brien 1890, 4).

5 The size classes for small, medium, and large holdings are < 2.4, 2.4–7.6, and > 7.6 hectares respectively (O'Brien 1890, 14).

6 Tenancy in Kangra is complicated. Historically some "tenants" held a hereditary right to cultivate land, were responsible for paying half the land revenue, and divided the remaining surplus with the "owner" (Barnes 1855, 19). At the other extreme were "tenants-at-will" who had no hereditary occupancy rights to the land they cultivated. Many tenants also owned land or were artisans or others whose trade was their main occupation. Lyall divides tenancies into thirteen different classes depending on the nature and origin of a tenant's claim to cultivate land (1874, 223–24). The range of payments tenants made to the landowner included (1) none, other than their share of the land revenue, (2) fixed cash rents, (3) fixed rents in kind, (4) part grain and part cash rents, and (5) shares ranging from less than one-fourth to more than one-half of the harvest. The majority of tenants paid half the produce as rent in Kangra in the late nineteenth century, and the average tenant's holding was 0.5 hectare (Lyall 1874, 228, 84). In-kind payments made at harvest time to the *kohli* (water mas-

ter), artisans, and others were set aside before the harvest was divided between the tenant and landowner (Lyall 1874, 59).

7 The Mahajan caste, composed primarily of traders, has the largest average landholding. During the British period, many Mahajan families accumulated much land and wealth as moneylenders. Before land reform legislation in the 1970s, twenty-five of the fifty-eight holdings greater than ten acres belonged to Mahajans, eighteen were owned by Brahmins, and nine were owned by Rajput families (Parry 1976, 56).

8 Landholders did not at first realize that the authority to grant such rights now rested with the community of landholders within each village, and no longer with the state. Lyall comments that during the operations of the second settlement, he "must have been asked several hundred times by landholders to give them *pattah,* or grants for waste plots" (Lyall 1874, 22).

9 Lyall provides more basis for his argument that the landholders could not be held responsible for the reclassification of uncultivated areas as joint village property because although they did adopt the term *shamilat* for unenclosed waste, they frequently referred to it using the contradictory term *shamilat sarkari* (government common property), thus indicating that the settlement notwithstanding, they still associated those areas with government ownership.

10 This was consistent with British policy encouraging agricultural expansion and conversion of forests to agricultural lands in the Ganga-Jamuna Doab during the preceding decades (M. Mann 1995, 211–12).

11 Landless households were allowed to "graze a few cattle, to collect dry wood, and to cut grass" in undemarcated areas, but only "as an act of grace and on sufferance," and only to meet domestic requirements (A. Anderson 1887, 6).

Vasant K. Saberwal

Environmental Alarm and Institutionalized Conservation in Himachal Pradesh, 1865–1994

The Gangetic watershed, in fact, has all the physical conditions present that must unleash powers of destruction, denudation, and desiccation, against which man is helpless, if the only real defense—i.e., natural vegetation—is once destroyed. . . . [The Gangetic watershed has an] enormous domestic animal population with intense and universal overgrazing of all waste and uncultivated land, against which the protective natural vegetation has no chance to survive or function. Although the threat of erosion to human property (and even existence) is now clearly recognized in all five continents of the globe, I doubt if there is any area in the world where the danger is so great as in the Gangetic basin, where so little is being done to meet the danger, and where a population of 100 million human beings is likely to be affected. (Smythies 1939, 179)

INTRODUCTION

The Malthusian specter of impending disaster, in the form of eroding hillsides, most typically contextualized in the case of Himalayan slopes slipping away, and advancing deserts, most often typified by the Saharan desert marching southward, has taken center stage in the degradation/desiccation discourse of the past century.[1] Within this discourse, Himalayan erosion has been attributed to the land use practices of an expanding farmer and agro-pastoralist population (Eckholm 1975; Myers 1986), and the south-ward extension of the Saharan desert to overgrazing by even faster grow-ing east African pastoralist communities (Brown 1971; Lamprey 1975, 1983; Eckholm 1977).

A growing body of literature is questioning both the scale and the mag-nitude of these and other descriptions of degradation, as well as the causal mechanisms that have been most commonly associated with these phe-nomena. Research within the east African rangelands suggests that tradi-tional herding practices may not be as degrading of the range as has been claimed in the past (Ellis and Swift 1988; Behnke and Scoones 1993). Re-searchers are pointing to fluctuations in the Saharan desert's boundaries in response to cyclical changes in rainfall intensity as a more accurate descrip-tor of change than a unidirectional southward movement linked to grow-ing pastoralist populations (Forse 1989; Binns 1990; Mace 1991; C. Tucker et al. 1991). Large-scale erosion in the Himalaya has similarly been linked more directly to naturally occurring high rates of erosion than to flawed subsistence practices of farming and pastoral communities (Hamilton 1987; Ives and Messerli 1989).

If "good science" does not underlie these premonitions of environmen-tal catastrophe, what then are the roots of such "myths of environmental degradation" (Forse 1989; Binns 1990), as they are increasingly referred to? Can this widespread phenomenon be attributed simply to bad science, or have political or cultural factors played a role in shaping images of degra-dation?

This essay posits an explanation for the origins of the alarmist degrada-tion discourse associated with the Punjab Himalaya. It is true that the gen-eral argument put forth by the Punjab Forest Department during the late nineteenth and early to mid–twentieth centuries regarding the relationship between human pressures and soil erosion, floods, and desertification was part of a wider scientific literature of the time. In addition, however, I argue here that sustained intragovernmental opposition to the conservation poli-cies of the Forest Department in effect "forced" the department to adopt a specific, highly alarmist position with regard to the environmental impact of unregulated use of forest lands.[2]

A number of scholars have recently explored the origins of concerns ar-ticulated by foresters during the colonial and postcolonial periods and the translation of these concerns to conservation policy. Ramachandra Guha

(1989) and Gadgil and Guha (1992) have posited that colonial and post-colonial forest conservation policies, with their emphasis on reducing local access to, and use of, forest resources, stemmed primarily from the imperial state's concern with maximizing forest revenues. Grove (1995), on the other hand, suggests that restrictive policies of the colonial government have their origins primarily in colonial forester concerns with the environmental consequences of extensive deforestation, consequences that were particularly noticeable within the outlying and somewhat delicate ecosystems of small islands. Rajan (1994) criticizes Guha's and Gadgil's alignment of forester and colonial interests, suggesting, as does Grove, that foresters brought their own environmental, rather than purely economic, concerns to the colonial state. Rajan goes on to criticize Grove for his characterization of forest policy as a primarily colonial experience, suggesting that foresters trained in the tradition of European forestry essentially transferred ideas and frameworks from the continent to the colonies.

In an important deviation from these studies of concerns and policies articulated at the level of the imperial government, Prasad (1994), Rangarajan (1996), and Sivaramakrishnan (1996) all point to the importance of the local context and the field experiences of foresters in the ultimate shaping of conservation policies. Sivaramakrishnan (1996) also points to the institutional context as a powerful factor in shaping the writings of colonial foresters, suggesting that the combination of field experience on the one hand, and institutional pressures and negotiation between local community and colonial government on the other, results in a dynamic policy formulation process, one beholden to neither economic nor ideological interests in the metropole, although both may play a role in the ultimate shaping of policy. My suggestion of the importance of the institutional context in shaping policy complements Sivaramakrishnan's work, whereas my focus on the rhetoric of conservation points to a directionality that is perhaps at odds with Sivaramakrishnan's suggestion of flux in light of experience.

Following this introduction is a description of the physical and social context within which this study is located. A third section, divided into three subsections, explores the changing rationale for better forest conservancy over the past century and a half, documenting the development of a highly alarmist discourse on degradation in the region. A fourth section documents the scientific backdrop to the department's changing rationale for forest conservancy, highlighting the absence of empirical research and the department's continued use of models of forest functioning that have been discredited within the scientific literature since the 1920s. I then briefly outline the dimensions of the intragovernmental resistance to Forest Department efforts to regulate access to forest resources over the past century and a half. A brief conclusion highlights the key arguments made in the paper.

PHYSICAL AND SOCIAL CONTEXT

For the most part, the material presented hereafter relates to the state of Himachal Pradesh.[3] A series of parallel mountain ranges critically influence the ecology of the region, with the low-lying Hoshiarpur Siwaliks to the south and the Great Himalayan range to the north bounding my area of focus. The Siwaliks are made up of soils that have not been subject to the geological compaction that soils of the main Himalayan ranges have. As a result, the soils of the Siwalik hills are loose and more susceptible to being eroded than the soils of the various Himalayan ranges to the north. In general, soil and vegetation characteristics of this mountainous state vary dramatically, depending on elevation, aspect, and slope. Vegetation changes from the scrub forest of the Siwaliks and lower Himalaya to oak and coniferous forests in the middle Himalaya, which give way at around 12,000 feet to alpine meadows of the Dhaula Dhar, Pir Panjal, and Great Himalayan ranges.

The region is subject to intense grazing pressures. Official statistics suggest that more than 5 million cows, goats, sheep, and buffalo grazed approximately 30,000 square kilometers of land in 1992 (HPFD 1993). This pressure is unevenly distributed, geographically and seasonally. Grazing pressures decrease as one moves from the low-lying Siwaliks to the alpine meadows of the high Himalaya. Differences in geology, vegetation, and grazing pressures have largely been ignored within the undifferentiated and sweeping condemnation of local land use practices by the Punjab and Himachal Forest Departments but are critical to our developing a more nuanced understanding of the relationship between forests, land use practices, and soil and water conservation in the region.

That the Forest Department attempted to restrict grazing within the Punjab forests in the mid–nineteenth century is not surprising, given that most foresters of the time had been trained in Europe. Within European forestry, grazing was simply seen as being incompatible with the production of timber. Although a blanket ban on grazing was later challenged by foresters within the context of Indian silviculture,[4] it is true enough that intense grazing pressures are generally incompatible with the maintenance of high-quality stands of timber.

What did change over time was the rationale put forth by the Forest Department regarding the need to maintain forest cover in the Punjab. Over time this changed from being a purely commercial argument—grazing and timber production could not coexist—to a more environmental argument that centered on the connections between forests on the one hand and climate and the hydrological balance on the other. This latter, more politically charged view has come to underlie the focus of the Forest Department today, and I argue here that the roots of this transition can be found within the intragovernmental resistance to Forest Department regulations.

1860s–1880s: An Economic Rationale

As with the rest of the country, and in line with a number of earlier analyses (R. Tucker 1988; Ramachandra Guha 1989b), in the mid–nineteenth century the Punjab Forest Department was primarily concerned with the extraction of timber and fuel from the Punjab. Both resources were in heavy demand owing to the ongoing extension of the northern railway line across the Punjab. Meeting this requirement, and the demand for timber from other parts of the country, provided the principal impetus to the earliest forest conservation within the Punjab.[5]

That timber and fuel production was the overriding concern of the Forest Department in the 1860s and 1870s is evident from a number of reports written at the time. For example, Dr. J. L. Stewart, officiating conservator of forests, submitted lengthy reports on the commercially valuable deodar *(Cedrus deodara)* tracts of the Ravi and Chenab Rivers, and a report on the fuel-bearing tracts of the Punjab.[6] During this period, a number of reports also referred to the dwindling fuel supplies of the Simla region and the army cantonments close by,[7] as well as that of the northern railway line.[8] These reports provide an essentially economic justification for improved forest conservancy, with the problem couched in terms of an impending fuel famine rather than potential or ongoing soil erosion or flooding—key elements of the desiccationist discourse taken up later in the century.

As Grove (1995) and Rajan (1994) have pointed out, an explicit focus on the climatic and protective capacities of forests had formed the crux of the justification put forth by conservationists for the establishment of the Forest Department in the early to mid–nineteenth century. And yet while making recommendations on the kinds of forests that should be conserved, foresters appeared to focus almost exclusively on the commercially valuable species of deodar, *sal (Shorea robusta), chil (Pinus longifolia),* and bamboo *(Dendrocalamus strictus),* suggesting a commercial rather than environmental interest in these forests.[9] If climatic and protective issues were of concern to these early foresters, one would have expected a greater push for the reservation of forests based on proximity to streams and the importance of particular watersheds rather than on the species that dominated particular forest tracts.

The Hoshiarpur Chos Act of 1900: The Shaping of a Discourse
The Hoshiarpur Chos Act of 1900 banned the grazing of goats and sheep in the Hoshiarpur Siwaliks. The act was justified solely on environmental grounds, with broad agreement among Revenue and Forest Department officials that intense grazing pressures had served to reduce vegetation cover, which in turn was responsible for high rates of soil erosion, as well as an increased intensity in the flooding of torrential streams that are a part of this landscape. Concerns about the Hoshiarpur Siwaliks had been voiced by many officials of both the Forest and Revenue Departments, and over many decades before the enactment of the Chos Act, and by its enactment, the Forest Department acquired the authority to buy out grazing rights within the Hoshiarpur Siwaliks.[10] In many ways, the Hoshiarpur Chos represented the first real success of the Forest Department in its attempts to reduce grazing pressures within the Punjab.

The highly erosive nature of the Siwalik hills played a key role in enabling the Punjab Forest Department to push through the Chos Act. While grazing pressures undoubtedly contribute to erosive processes in these hills, even within the Siwaliks there are obvious differences in erosivity from one site to the next, a function most specifically of slope and aspect (Puri 1949). Separating the respective roles played by grazing and natural processes, however, in identifying causation with regard to Siwalik erosion is a difficult task, one rarely attempted by foresters past or present.

The fact remains that visually parts of the Siwalik landscape bear the appearance of being in a highly degraded condition. Crumbling, exposed soil in shallow ravines typifies parts of the range. Another feature of these hills is the occurrence of flash floods during the monsoons, when a large amount of debris, including boulders and rocks, is washed down boulder-strewn streambeds, locally referred to as *chos*. These streams empty into the densely settled Punjab plains, thereby providing an immediacy to the situation that is not true of the Himalayan ranges to the north.

In the mid- to late nineteenth century, the Punjab Forest Department associated large-scale destruction with these streams. The action of the streams was thought to lead to a cutting away of cultivated fields in the hills, leading to large-scale loss of property; cultivation in the fertile Punjab plains in a fifty-kilometer-wide belt south of the Siwalik hills was considered to be threatened as a result of soil deposition by these streams; and the cities of Hoshiarpur and Jullundur were considered under the perennial threat of being covered by the debris brought down by these streams. All this trouble could be avoided if goat and sheep grazing were to be prevented within the Hoshiarpur Siwaliks.

Although there is a definite truth to the Forest Department's assertion regarding the need for better-regulated grazing pressures in the Siwaliks, the continued existence of Hoshiarpur and Jullundur cities, and the fertile tract of agricultural land to the south of the Hoshiarpur Siwaliks, despite an increasing rather than decreasing cattle and human population over the past century, suggests a certain exaggeration in late-nineteenth-century forester descriptions of the chos problem.

Hyperbole was not new to the Forest Department. In the past it had used threats of timber and fuel famines to press for more strict forest conservation, without managing to effect a complete ban on grazing in other parts of the Punjab Himalaya. What was different with regard to the Hoshiarpur Siwaliks was the imagery of devastation that the department was now able to work with. All the apparently obvious ingredients for large-scale degradation—exposed soil, mountainous terrain, and boulder-carrying torrents—were present in the Siwaliks, and as a result the Forest

Department was able to generate much greater support for the restrictions it proposed.

The success with which the Forest Department focused governmental attention on the chos problem during the 1890s coincided with a growing effort to introduce regulations in other parts of the state as well. The Kulu and Kangra forest settlements were completed in 1894 and 1897 respectively and for the first time introduced measures aimed at regulating pastoralist use of grazing resources in the state. The sheep- and goat-herding Gaddi, for example, were now required to move at least five miles a day; a fine was to be imposed on any herder who stayed longer than one day in any location; and grazing fees were imposed on these herders for the first time in British rule. In 1914 two additional taxes—a cattle tax on the resident cultivators, and a grazing tax on the Gaddi—were introduced in an attempt to induce a reduction in the numbers of animals maintained by these two communities.

But even as tighter regulations were being introduced in an attempt to reduce grazing pressures in the state, and as grazing came to be labeled the principal agent of destruction within the Punjab forests (see Annual Progress Reports of the Punjab Forest Department, 1904–1914), there was a general recognition that geologic characteristics of the grazed substrate critically moderated the impact of intense grazing pressures. Speaking of overgrazing in Kangra district, Deputy Commissioner Fagan stated that "the general constitution of the soil prevalent in the tracts concerned is not, it appears to me, as a rule, such as to favor rapid erosion. This feature of the situation tends to limit the distance to which damage consequent on denudation can extend from its source and to discriminate the case of Kangra from that of the Hoshiarpur Siwaliks, the main component of which is soft, friable sandstone."[11]

That such distinctions were made by some foresters as well can be seen in a note by Assistant Conservator of Forests Holland on the condition of the lower slopes of the Himalaya and the Siwalik hills.

The lower slopes of the Himalayas are well-wooded with scrub and bamboo jungle, and the rock which occurs is harder than the sandstone of the Siwaliks. Erosion is very slight indeed, and is only met with locally along the foot of the hills where its action is very slow. The Siwaliks as a whole present the appearance of a mass of ravines, peaks, cliffs, scarps and bare slopes densely crowded together of varying shades of brown, for vegetation is very scarce indeed.[12]

Such distinctions between conditions in the Siwaliks and other parts of the Kangra district, or even between different parts of the high Himalaya, were occasionally made by foresters during the 1920s and 1930s.[13] Certainly there was no unified, undifferentiated view within the forester community of the impact of grazing on the landscape. As I will try to demonstrate, this diversity of opinions gives way over time to a more simplistic and unified understanding, or portrayal, of the impact of grazing on the Kangra landscape.[14]

1920–1959: A Nondiscriminating Discourse

In 1911, Kangra district commissioner Colonel Powney Thompson suggested that the "degraded" condition of the Kangra forests was due to lack of local support for the Forest Department, and the unpopularity of its forest conservation measures. As a means of generating support for forest conservation, Thompson suggested that local communities be given greater control over these lands. As a first step in that direction, he recommended that unclassed forests be transferred back to the control of the Revenue Department.[15] By 1919, despite considerable opposition of the Forest Department, almost half the forest lands in the state were transferred to the control of the Revenue Department.[16]

Following this transfer, the conservator of forests, J. W. Grieve, put forward a proposal to solve the grazing problems of the villagers.[17] The proposal contained detailed suggestions of a scheme by which hay, cut from forests now retransferred to the control of the Forest Department, could be transported over large distances using telegraph wires and used to feed livestock corralled in specific areas rather than being allowed to graze freely in state forests. Although the scheme was not considered seriously by the government, the absurdity of carting hay over a network of telegraph wires crisscrossing the Punjab is testimony to the lengths to which the department was willing to go to secure control over these lands.

More important, however, Grieve's report contains a description of the Kangra forests (not the Hoshiarpur Siwaliks) that is in stark contrast to the comments made by Fagan and Holland quoted earlier. Indeed, his description of these lower hills appears to suggest the imminence of the disappearance of the forests, the soil these forests grew in, and eventually the herders themselves.[18] That animal and human pressures have only increased since the 1920s, and a profitable herding enterprise has continued to this day, is testimony, once again, to the excessively alarmist nature of Grieve's analysis. Smythies's comment at the start of this chapter exemplifies the general tone of much forester writing of the 1930s and 1940s.[19]

The forester agenda for having a greater say in how forests were used and managed received a dramatic fillip from the American Dust Bowl. The international media relayed stories of New York City being blanketed by Nebraskan soil, and news reports indicated massive losses of topsoil. Paul Sears's book *Deserts on the March* appeared in 1935, coinciding with the growing concern in Australia, Africa, India, and the United States about the desiccating consequences of deforestation and the mismanagement of land resources. Beinart (1984) and D. Anderson (1984) point out that the American Dust Bowl served to fuel concerns about soil erosion resulting from the misuse of land resources in South Africa and East Africa, respectively. The Indian Forest Service both fed this global concern and, in turn, used it to justify the formulation of more restrictive policies.

That the rhetoric of the 1930s and 1940s has sustained to this day is demonstrated by a policy document written by Divisional Forest Officer B. S. Parmar (1959, 5), one that continues to be used as the basis for Himachal's grazing policy. Parmar describes the condition of the Himachal forests in familiar, generalized terms:

> Uncontrolled and excessive grazing in village waste lands adjoining habitations and cultivation have rendered them unstable and eroding. Hundreds of slips can be seen everywhere. A glance at any scrub-covered hill, not protected from continuous and almost always excessive grazing, is sufficient to show the steady deterioration of the vegetation, which in many cases must disappear eventually. . . . The lower slopes of the ban oak forests are almost bare of grass and the soil is covered with stones, while erosion accompanied by land slips, continues apace.

In 1986, well-known environmentalist Norman Myers stated that "primarily because of deforestation in their headwater regions, the [Indo-Gangetic] river systems are increasingly subject to disruption, leading to floods followed by droughts. . . . All in all, these plains have been described as the 'greatest single ecological hazard on Earth'" (Myers 1986, 64). The conviction of the assertion brooks little dissent, even though the simplistic relationship Myers makes between forests, flooding, and drought has been in disrepute for the past seven decades.

FOREST DEPARTMENT ALARM IN THE CONTEXT
OF SCIENTIFIC INQUIRY

The alarmist position of the Forest Department is particularly interesting in light of three phenomena. First, coincident with the increasingly alarmist rhetoric adopted by the Forest Department from the 1930s on, is the acceleration in the extraction of timber from Himachal's forests, with commercial extraction increasing from 1.5 million cubic feet at the end of the 1920s to more than 20 million cubic feet by the 1970s (Saberwal, forthcoming). There is thus a dramatic disjunction between the Forest Department's stated concern regarding deforestation and its own role in decreasing forest cover in the region.

Second, with regard to the Punjab, is the absence of empirical research supporting the proposition that grazing pressures were leading to a rapid depletion of vegetation cover. Over the course of the past 150 years, there is no experimental manipulation to evaluate the impact of grazing on the state's forests; there are no long-term data regarding the permanence of change in species composition, declining vegetation cover, and increasing soil erosion. There is also no attempt to differentiate erosion resulting from natural processes and that resulting from human activities, a key problem in identifying causation with regard to land degradation in the Himalaya (Hamilton 1987; Ives and Messerli 1989; CSE 1991). What evidence has been presented has been anecdotal, with localized instances of landslips being used to portray a statewide problem of degradation and impending environmental catastrophe.

Third, the Forest Department continues to use a model of forest functioning that has been discredited within the ecological literature since the 1920s.[20] By the late nineteenth century, serious problems were being raised within and outside the international forester community regarding the accuracy of the model of forest functioning in vogue at the time. By the 1920s, long-term empirical research in Switzerland, Colorado, Russia, and France, and a considerable amount of research since (particularly Bormann and Likens 1979), was demonstrating that the removal of forests need not, in fact lead to alternating cycles of drought and flooding. Instead, and directly contradicting one of the most cherished precepts of forestry, research has demonstrated time and again that the removal of forests can lead to a *year-round increase* in water flow, owing to the reduction in transpirational losses associated with dense forest cover.[21] Despite this evidence to the contrary, official policy and popular writings continue to be premised on the notion that deforestation leads to an intensification of flooding during the monsoon and diminished water supplies during the dry season.

SUSTAINED RESISTANCE TO FOREST
DEPARTMENT CONTROLS

The formation of the Indian Forest Department in 1865 initiated a power struggle between the Forest and Revenue Departments, one that continued to simmer well into the middle of the twentieth century. A number of researchers have pointed to the existence of this conflict, which essentially revolved around the Forest Department's attempts to secure greater control over the management of forest lands in the face of determined resistance by the Revenue Department (Ramachandra Guha 1990a; Rajan 1994; Rangarajan 1996; Sivaramakrishnan 1996).[22]

Within the Punjab, this conflict over control took many forms, including contestation over the primacy of departmental control over forest lands as well as over the formal authority and status of Forest Department officials vis-à-vis their counterparts in the Revenue Department. During the first few decades of the conflict, the Revenue Department successfully campaigned against large-scale transfers of land to the Forest Department on the grounds that the latter lacked adequate staff to effectively manage these lands.[23] The Revenue Department also suggested that the transfer of all forests to the control of the Forest Department, in a country where locals were so dependent on forest products for their survival, would greatly diminish the status and authority of the Revenue Department, thereby reducing the district commissioner's ability to adequately control the district.[24] The Revenue Department also predicted that a strict enforcement of Forest Department regulations would lead to discontent among local residents, and eventually to political turmoil. As a means of curtailing such discontent, the Revenue Department argued that the district commissioner, the ranking Revenue Department official in a district, be given discretionary powers to relax Forest Department restrictions to accommodate local interests, particularly during times of unforeseen hardship. This authority was formally acquired by the Revenue Department in 1884, owing to the passage of legislation that specifically subordinated the district forest officer to the control of the district commissioner.[25]

The conflict between these two departments continued into the 1930s. One of the most dramatic examples of the continuing tension between the departments surfaced in 1913, when the district commissioner of Kangra, Colonel Powney Thompson, suggested that the forests of Kangra district were deteriorating owing to the lack of local support for the unpopular policies of the Forest Department. He recommended that the majority of forest lands that had been transferred from the Revenue Department's control to that of the Forest Department in the late nineteenth century be re-transferred back to the control of the Revenue Department.[26] In 1919 more than half the forested lands of Kangra district were transferred to the control of the Revenue Department; predictably, the Forest Department opposed the transfer, with conservator of forests R. McIntosh stating that "in the absence of scientific treatment [the transferred forests] must inevitably degrade."[27]

I lack the space to provide a detailed account of the frictions between the two departments.[28] Interestingly, Ravi Rajan (1994) points out that Forest Department policies advocating curbs on resource extraction were met with general resistance throughout the British empire well into the middle of the twentieth century. The depression led to a series of cutbacks in government spending on a variety of issues; forestry was one of the hardest hit. It is likely that foresters were forced to respond to budgetary cutbacks by the adoption of an increasingly alarmist rhetoric with regard to deforestation and the consequences thereof.[29]

The conflict between the Forest and Revenue Departments in postindependence Himachal Pradesh is less evident than in the earlier decades of the twentieth century. Even so, the Forest Department continues to experience a great deal of difficulty in the implementation of its most restrictive policies, owing primarily to politician interference in Forest Department management. Between the 1960s and the 1990s, senior politicians, including the chief minister, the forest minister, and the Speaker of the Legislative Assembly, have interfered in the functioning of the Himachal Pradesh

Forest Department, instructing officials to accommodate herders in grazing areas, to open forests formerly closed to grazing, to relax regulations for extended periods of time, and so on.[30] Herders inform me that few, if any, grazing regulations are enforced today.[31]

CONCLUSION

We observe, then, a gradual change in the rationale for forest conservancy, a change unsubstantiated by research and discredited within the scientific literature for the better part of the twentieth century. Simultaneously we observe an institutional dynamic in which the Forest Department has been frustrated over the past century and a half in its attempts to gain greater control over forest lands in the Punjab Himalaya. I have argued that this institutional resistance has, in effect, forced the Forest Department to adopt an increasingly alarmist tone to its predictions of environmental disaster, if only to justify its claims regarding the need for better forest conservation.

It is not my intention to suggest an absence of degradation in the region. On the contrary, the removal of forests has obvious and serious implications with regard to local requirements for fuelwood, grazing, and other nontimber forest products. My commentary, rather, is on the sweeping and alarmist nature of Forest Department generalizations. Such generalized descriptions of degradation serve to obscure both the nature of the problem and potential solutions.

There is considerable uncertainty with regard to our understanding of ecological phenomena. Climate, topography, and geology, among other things, as well as human land usage, help shape the vegetation of a given region. And there are always complex interactions, at various levels, that serve to further complicate our understanding of ecological processes. Given the opposition that the Forest Department has had to deal with since its establishment as a government department, it is hardly surprising that foresters have been less than vocal about the inexact nature of ecology as a science.[32] In effect, the Forest Department has historically lacked the institutional space to acknowledge the uncertainty inherent to our understanding of large-scale ecological phenomena such as floods, desertification, and so forth, and the difficulties of establishing causality with regard to these phenomena. Such an acknowledgment would necessarily undermine the position of the Forest Department in the face of the considerable intragovernmental resistance to conservation policies.

Over time, we observe a petrification of the simplistic position that overgrazing is rampant within Himachal Pradesh, and the assumption that overgrazing and the resultant deforestation are leading to an intensification of flooding within the Indo-Gangetic plains, a gradual drying of the area, and a rapid reduction in the region's capacity to continue supporting current levels of resource extraction. This attribution of almost magical protective capabilities to forests, and the benefits to populations downstream of these mountain forests, has served as the primary justification in Forest Department efforts to exclude pastoralists from Himachal's state forests, a position that is problematic not only for the livelihood problems it poses for herders but also for the inappropriate handling of flooding in the Indo-Gangetic plains.

NOTES

I thank the staff at the Himachal Pradesh State Archives, the National Archives, and the India Office Library for assistance in tracking sources. Grateful thanks to K. Sivaramakrishnan for encouraging me toward such an analysis. I thank participants of the Agrarian Studies Graduate Student Colloquium (spring 1995), as well as those of the Agrarian Environments workshop (May 1997), both at Yale University. Jesse Ribot, Nancy Peluso, Ruhi Grover, and James Scott commented on various drafts of this paper. The study was funded by the Biodiversity Support Program (a consortium of the World Wildlife Fund, the Nature Conservancy, and the World Resources Institute, with funding by the U.S. Agency for International Development), the Yale Center for International and Area Studies, the Agrarian Studies Program at Yale University, and the American Institute of Indian Studies. The current version of this chapter was finalized in 1996 and 1997 while I was a Pacific Basin Research Fellow of Soka University. During this time I was provided logistical and administrative support at the Harvard Center for Population and Development Studies, for which I am grateful. The opinions expressed here are those of the author and do not necessarily reflect the views of the people or institutions I have mentioned. An earlier version of this essay was published as "Bureaucratic Agendas and Conservation Policy in Himachal Pradesh, 1865–1944," in *Indian Economic and Social History Review* 34 (4): 465–98, 1997.

Abbreviations

CF = conservator of forests; Comm. = commissioner; C&S = commissioner and superintendent; DC = district commissioner; Dec. = December; DFO = divisional forest officer; Div. = division; FC = financial commissioner; Feb. = February; Govt. = government; HSA = Himachal State Archives; IF = *Indian Forester;* IG = inspector general; Jan. = January; Offg. = officiating; Progs. = proceedings; PWD = Public Works Department; RAC = Revenue, Agriculture, and Commerce; RAD = Revenue and Agriculture Department; Secy. = secretary; Sen. = senior; Sep. = September.

1 I use the term "desiccationist discourse" here to incorporate a specific, interconnected body of ideas, centered on the connections between deforestation on the one hand and increased erosion, flooding, and overall aridity on the other. Essentially, the discourse or narrative suggests that unregulated grazing by rapidly growing goat, sheep, and buffalo populations is leading to a rapid removal of vegetation cover, which in turn is responsible for increased soil erosion, enhanced water runoff (and therefore flooding during the rains), and, owing to an ineffective recharging of underground aquifers, prolonged drought during the dry season. Alternating cycles of drought and flooding in the Indo-Gangetic plains, with consequences for the 400 million residents of these plains, are seen as a direct result of rampant overgrazing by pastoralist communities. A key characteristic of this discourse is the simplification of ecological phenomena, one that enables a smoothing over of variations from one context to another. This discourse has been adopted both within the popular media and within official conservation circles in most parts of the world, although my primary reference here is to the discourse as it matured within the writings of the Punjab and Himachal Forest Departments between 1865 and 1994.

2 In contrast to my focus on the institutional dynamic within which state policies are articulated, Mark Baker in this volume focuses primarily on changes within civil society, particularly in the context of the state's attempt to transform property rights in Kangra district of the Punjab.

3 The state of Himachal Pradesh was initially constituted following its separation from the Punjab at the time of Indian independence and acquired its current geographic boundaries in 1966.

4 See, for example, G. S. Hart, IG Forests, note on a tour of inspection in the Kulu and Kangra Forest Divs., Punjab. Dec. 1915. See also Sivaramakrishnan 1996 for an account of changing forester attitudes toward grazing in the forests of eastern India.

5 See note by governor general of India, following a tour of the Punjab plains. Foreign Dept., A Progs. March 1851, nos. 67–68.

6 General Report on the Deodar Forests upon the Upper Chenab in Chota Lahaul and Pangi; General Report on the Deodar Forests of the Ravi; Report on the Fuel Bearing Tracts of the Southern Punjab. Punjab PWD (Forests) A Progs., July 1866, nos. 37, 48.

7 See, for example, B. H. Baden-Powell, CF, Punjab, writing to the Offg. Secy. to Govt. Punjab, 5 March 1872, regarding the need to stop all grazing in forests close to Kassauli immediately; correspondence regarding "Supply of Fuel to Simla Hill Stations," Punjab Forest A

Progs., Sep. 1875, no. 3; W. Schlich, CF, Punjab to Secy. to Govt. Punjab, 1 March 1880. "Plantation of Hill Cantonments, Dagshai." Punjab Forest A Progs., RAD, April 1880, no. 2.

8 See, for example, remarks by Under Secy., Railways and Forest Branch, PWD NW Provinces. RAD (Forests), B Progs., March 1865, nos. 32–36; Report on the Supply of Fuel for the Punjab Railway, by Dr. J. L. Stewart, 17 Sept. 1864.

9 See, for example, H. Cleghorn, 1861, Report upon the Forests of the Punjab and the Western Himalaya. Roorke, Memorandum on grazing in government forests and waste-lands. B. Ribbentrop, CF Punjab. 7 Dec. 1883. Basta 27, Serial 407, File 10 (127), HSA, p. 4.

10 RAD (Forests), May 1899, A Progs., nos. 7–10.

11 Punjab Forest A Progs., April 1914, no. 18.

12 Report on Denudation in the Siwaliks and Lower Himalayas between the Sutlej and Ghaggar Rivers, 7 June 1912, Punjab Forest A Progs., July 1912, no. 6.

13 For example, see the following notes as examples of foresters countering suggestions of grazing-induced degradation in the high mountains of the Uhl Valley, just south of Bara Bangahal. S. Deans, deputy CF, Kangra, n.d. Serial 309, Basta 20, File 10(34)-II, HSA; DFO Kangra, to Glover, CF, Punjab, 25 June 1937, Serial 309, Basta 20, File 10 (34)-II, HSA.

14 For a more detailed account of this transition see Saberwal 1997.

15 P. Thompson, DC Kangra, to Comm., Jullundur Div., Dec. 30, 1911. Punjab Forest A Progs., April 1913, no. 16.

16 Sen. Secy. to the FC, Punjab, to the Secy. to the Govt., Punjab, 12 Feb. 1920. Punjab Forest A Progs., May 1920, no. 4.

17 Note on the economics of nomadic grazing as practiced in the Kangra district. From the CF Eastern Circle, Punjab to the FC, Punjab, 1–2 March 1920. Basta 24, Serial 352, HSA Simla.

18 J. W. A. Grieve 1920. Note on the economics of nomadic grazing as practiced in Kangra district. *IF* 28 : 333.

19 For additional examples of the dramatic increase in the level of alarm in forester writings at this time, see writings in the *Indian Forester,* also Saberwal 1999.

20 It is not that there is a "scientific" literature that accepts the alarmist discourse on degradation. There is generalized writing that supports the position, but writing that is not "scientific" in terms of an examination of the specific components of the overall discourse. Thus there is no "scientific" work that examines the specific relationship between flooding and vegetation cover in the Himalaya, even though there is a large literature that takes this as established fact. My reference to a scientific literature that discredits the alarmist discourse is to scientific literature that has examined the specifics of the relationship between forest and vegetation cover and the dominant components of the alarmist discourse—flooding, desertification, and soil erosion. For the most part, this work has not, in fact, taken place in the Himalaya—rather, it is empirical work, mainly in the United States, that discredits the simplistic relationship assumed by the alarmist rhetoric (see Busch and Hewlett 1982 for more details).

21 See Busch and Hewlett 1982 for a review of research on the role of vegetation change in shaping catchment hydrological regimes.

22 Researchers differ in their evaluation of the duration of this conflict. Guha and Rangarajan have suggested that the passage of the Indian Forest Act of 1878 effectively evened the

power differentials between the two departments, whereas Rajan and Sivaramakrishnan have pointed to a longer-lasting conflict. This chapter supports the latter view.

23 See, for example, FC Punjab to the Secy. to Govt. Punjab, PWD, 27 July 1865. PWD, A Progs., Feb. 1866, no. 18.

24 E. H. Paske, DC Kangra to C&S Jullundur Div., 5 Oct. 1875. RAC (Forests) A Progs., Oct. 1877.

25 Punjab Forest A Progs., May 1885, no. 10. Rajan (1994) points out that the question of the subordination of Forest Department officials to their Revenue Department counterparts remained a central concern of foresters in India and other parts of the British empire even in the 1940s.

26 Powney Thompson to Commissioner, Jullundur Div., 30 Dec. 1911. Punjab RAD (Forests) A Progs., April 1913, no. 17.

27 McIntosh, CF Eastern Circle, Punjab, to the Sen. Secy. to the FC Punjab, 13–15 Jan. 1919. Punjab RAD (Forests) A Progs., Dec. 1919, no. 90.

28 See Saberwal 1997 for a more detailed account of this conflict.

29 In line with such deconstruction of the rhetoric of the Forest Department there is a similar need for deconstructing the language and rhetoric of the Revenue Department or other government agencies that may use "local" issues to their own institutional advantage.

30 I lack the space here to enter into a prolonged discussion of the relationship between the Forest Department and politicians of the state of Himachal Pradesh, or to examine the manner in which the growing political clout of herders and resident cultivators has undermined the Forest Department's capacity to regulate use of state forest resources. See Saberwal 1999 for a detailed analysis of the same.

31 Shubhra Gururani discusses the question of the state's attempts to restrict local access to state forests, and the manner in which villagers manage to circumvent these restrictions. In contrast to Gururani's description of the constant tussle between the forest guard and village women over questions of access to forest resources, the herders I worked with appear to have mobilized around ethnicity and occupation to generate substantial political clout, which they have used in circumventing the restrictions of the Forest Department (Saberwal 1999).

32 See Hilborn and Ludwig 1993 for a recent commentary from within the ecological sciences on the uncertainties in ecology.

State Power and Agricultural Transformation in Tamil Nadu

India is a developing country. This statement appears self-evident and sparks no real controversy. Through the lens of the post–World War II discourse of development, India is perceived primarily in terms of what it is not and, in turn, requires interventions designed to transform it into what it should be. Among the most prominent of India's development efforts have been schemes to transform its agrarian environments, both the natural world of land, trees, and water and the complex of rural social relations that physically and culturally shape it. These efforts have included the expansion of irrigation, promotion of new productive technologies (such as high-yielding varieties of food grains, chemical fertilizers, and pesticides), and increased government reach into every aspect of agricultural practice.

In recent years, critiques of international economic development have begun to explore the power of development discourse to produce authoritative truths about the so-called Third World (Escobar 1995; Ferguson 1990; Sachs 1992). Drawing on the work of Foucault (1979, 1980) on knowledge and power, and of Said (1979) on Orientalism, these analyses interrogate a regime of representation in which certain ways of being and thinking are devalued as "underdeveloped" while others are privileged as progressive. In doing so, they help to reveal how dominant agricultural development paradigms rest on particular constructions of what it means for natural landscapes and their human inhabitants to be productive, and converge with persistent popular views of these paradigms as environmentally unsustainable and culturally homogenizing. In India, critics have argued that state-sponsored development programs serve to limit diversity and concentrate power in a centralized state apparatus (Nandy 1989), and that the instrumental ideology of modern science adopted by the state leads to ecological destruction and human oppression (Nandy 1987; Shiva 1989b).

Although such critiques have been effective in mobilizing political opposition to environmental degradation and infringements of human and cultural rights, their acceptance of such oppositional categories as "state versus community" and "Western science versus indigenous knowledge" also limits their analytical power in the study of specific sites and processes of agrarian transformation. As several of the chapters in this volume argue, neither "states" nor "communities" may be assumed to be bounded and internally coherent. Furthermore, by focusing on the level of discourse or policy, these critical approaches tend to overvalue the directive power of "the state" (including its capacity to promote "science") and undervalue the agency of so-called beneficiaries to mediate the meanings and practices promoted by development programs.[1] A growing body of literature addresses the issue of local agency, and the processes by which colonial and postcolonial visions have been accepted, rejected, or reworked by their intended subjects.[2] Considerably less attention has been devoted to the everyday practices of the state as it strives to produce effects.[3]

In this chapter, I undertake an ethnographic study of a district Department of Agriculture to examine the related questions of the coherence, boundedness, and effectivity of the state in inducing transformation of the agrarian environment of rural Tamil Nadu.[4] Rather than assume that the state exists as a unified and autonomous entity with the power to effect development, I draw attention to the discourses and institutional practices that seek to achieve this power. My analysis is directed at the level of local officials for three related reasons. First, these officials are themselves rural social actors whose actions and aspirations are important to our understanding of agrarian and environmental change. Second, the position of local officials provides insight into the question of the state's coherence. As Gupta (1995, 376) notes, the lowest levels of a bureaucratic hierarchy are the point at which most people come into contact with the state. Thus the understandings and motivations of government officials at these local levels crucially mediate how the policies formulated at higher levels are actually presented to their intended beneficiaries. These understandings and motivations cannot be read from discourses or plans but require an approach that treats these officials as agents in their own right (Comaroff and Comaroff 1991, 9). Finally, the position of local officials is relevant to the question of the boundedness of the state. As Timothy Mitchell (1991) notes, it has always been difficult to define the limit of the state, the point at which its autonomy from "society" can be located and proclaimed. In the study of rural development, this line of demarcation is especially difficult to find when dealing with local officials, who often share significant social, political, and conceptual commonalities with agriculturalists.

A full picture of the institutional environment presented to farmers engaged in rice production would include state agencies concerned with land registration, collection of revenue, licensing of input traders, credit, marketing, and irrigation. However, I have concentrated on agricultural officials, and especially on extension workers, who make daily visits to instruct farmers in new technologies. Under the Training and Visit (T&V) system, which was implemented in Tamil Nadu in 1981 with World Bank funding, each district office of the state Department of Agriculture is headed by a joint director of agriculture (JD). Districts are then divided into agricultural divisions, each headed by an assistant director of agriculture (ADA), and further subdivided into blocks. Agricultural development officers (ADOs) in these block offices supervise a corps of village extension officers, known as assistant agricultural officers (AAOs), who are assigned to a regular circuit of village visits.[5]

These village extension workers, in particular, occupy a "blurred boundary" (Gupta 1995, 384) between the government office that employs them and the rural communities to which they are assigned. Almost all of them are from farm families and are familiar with the conditions and constraints under which farmers work. Second, most village-level workers—in contrast to their superiors in the block office and on up—are not B.Sc. graduates in agriculture. This, along with their farm family backgrounds, means that village-level workers share many agronomic concepts and terminologies with farmers. Third, though extensionists are not permitted to work in their home villages, they share a range of social ties with various segments of the population in their assigned villages. Although these same commonalities with farmers may characterize district officials at higher levels, they tend to be minimized by their educational backgrounds and professional training and the absence of a dialogic pull of frequent contacts with villagers. Village-level officials, by contrast, must orient themselves continually to the perspectives of farmers to maintain local credibility and intelligibility.

In occupying this blurred boundary, the role of local officials is also ambiguous with regard to the function of the state to promote development. Here, the state/society distinction becomes a distinction between those who are agents of development and those who are its targets. Although local officials seek to appropriate the status of state culture by casting themselves as agents of development, this positioning is unstable, as they are also seen as targets of development by their superiors. In this way, the operations of the state in transforming its rural landscape and citizenry are paralleled by efforts to create an effective workforce for the bureaucracy. Indeed, as I argue in this chapter, efforts to exert greater productive control over the agrarian environment become inflected by concerns with control over actors at lower levels of the state. The result is a shift toward greater rigidity in the everyday practices of the state, as well as the mediation and extension of development aims through the efforts of local officials to render them into terms meaningful to farmers.

LOCAL OFFICIALS AS AGENTS OF DEVELOPMENT

Generally speaking, local officials actively affiliate themselves with the aims and ideologies of development. These efforts at self-positioning must be understood in terms of the extent to which concepts of development and modernization have become incorporated into the construction of local identities and categories of social difference. Changes associated with development—including new commodities, technologies, systems of knowledge, and forms of work—are seen generally to be continuous with progress, with becoming "civilized" or "modern" *(nahariham),* and, in more personal terms, with aspirations for self-betterment. These ideologies have become an integral part of the ways that people see themselves and others in their society.

Gupta and Ferguson (1992) have cautioned against the tendency of scholars to imagine encounters between bounded and discontinuous cultural entities (such as "the West" or "the state" and "indigenous communities") and suggest that we become "interested less in establishing a dialogic relation between geographically distinct societies than in exploring the production of difference in a world of culturally, socially and economically interconnected and interdependent spaces" (14). In my view, development must be analyzed as one such differentiating force. Ideologies of development are themselves appropriated and assimilated into existing orders of meaning and social difference. Yet at the same time, ideas about development become new criteria of distinction, providing tools for the assignment of new identities to persons and places.

Stacy Pigg (1992) traces such a process in her study of the production of the social category of "the village" in the context of development programs in Nepal. The village, she argues, is defined by its relative lack of *bikas* (development), as this concept has come to construct differences in Nepali society along a scale of social progress. In Tamil Nadu, the word *nahariham* (most often translated into English as "civilization") is used in very similar ways. Because *nahariham* is related to *nagaram* (town, or city), the word suggests urban refinement and has probably long marked urban/rural differences. However, in contemporary Tamil Nadu, these differences are constituted in large part by the discourses and practices of development. As a result, many people aspire to attain the attributes of development, or—even better—to become agents of development. Pigg notes that "increasingly, the apparatus of *bikas* (the burgeoning of office jobs, the money brought in by foreign aid, the positions of influence in the bureaucracy) is the source of power, wealth, and upward social mobility" (1992, 511). As development programs often do not achieve their stated objectives, being an agent of development may well be more advantageous, in material terms, than being a target of development. In addition, a position as an agent of development brings with it the social prestige associated with being "progressive" and "modern." Finally, as development is relative, a job in a development bureaucracy also allows one to set up a constant contrast with those who are less developed.

It is certainly the case that many agriculturalists in Tamil Nadu look to various government agencies less for the services they provide than for the opportunity of joining them to become salaried workers. For sons of farm families, the Agriculture Department has seemed a particularly accessible avenue to this status. For example, one senior official who joined the department in the 1950s described how when he was still in school, he used to watch an older man pass by every day. The man drew his attention because he had a very nice manner, was well dressed and groomed, and rode a nice bicycle. He followed the man home one day and saw a plaque on his house reading "B.Sc., Ag." At that point, the official explained, he decided to apply to study agriculture in college, thinking, "I am also from an agricultural family."

In addition to material goods and professional advancement, another attraction of development that frequently crops up in conversations with both farmers and officials is the social power and prestige derived from familiarity with bureaucratic settings, advanced technologies, and new forms of authoritative knowledge. As one *grama sevak* (village worker) explained, "I wanted to become a grama sevak because the GS was powerful in the village. He knew about all the development activities and got credit for improvements." He went on to tell a story of how his knowledge and official position had won him respect in the village where he worked:

> One day a man came from next door with a bottle of pesticide. A new mother had accidentally drunk it thinking that it was neem oil, then got dizzy, and passed out. I went on my cycle to call the PHC ambulance, and then rode with her to the hospital. When we arrived there, the doctor would only let me—among all the relatives—enter the hospital room, because I was the GS. They pumped her stomach. She came back a few days later, and ever after that they invited me first to any wedding, or family function. Whenever I passed, she would stand. And whatever they got from their own field, they gave first to our house.

Another way that development workers signal their alignment with the ideology of development is their participation in the discourse of the traditional farmer. By this I mean the frequent references to farmers as being stuck in traditional ways of thinking and doing, unresponsive to incentives for change, and not aware of or not understanding the benefits to themselves of adopting the practices promulgated by the agents of development. This discourse is pervasive among development workers and reveals several characteristics of the development process. First, it marks the existence of a disjuncture between the practices of agriculturalists and an abstract ideal that is determined outside the context of local terms, understandings, and preferences to be the appropriate goal. Second, it indicates the existence of resistance to, or at least the absence of a ready embrace of, the new on the part of farmers. Finally, the image of the traditional farmer lays the ground for the operation of power. Farmers are traditional in contrast to something else that they "should" be—that is, modern or progressive. If value resides in the model and not in local practices, then induced change is justified as being for the beneficiaries' own good. In the case of agriculture, much of the disjuncture between development officials and farmers hinges on their images of the environment; as a result, farmers' knowledge of the environment and the actions they take in constructing it become the focus of induced change.

Compulsion in development is most frequently associated with large-scale infrastructure projects such as dams or centralized irrigation systems, which are physically cut into the landscape. Indeed, agriculture workers in Tamil Nadu refer enviously to the local Irrigation Department for its resources and its ability to exert direct control over the environment—and, by extension, farmers' behavior—without depending on the arduous process of changing farmers' minds. They point to this contrast as evidence of their relative incapacity to force farmers to follow their recommendations. Instead, extension officers describe themselves as "teachers" or "doctors," who try to enlighten or prescribe but do not force. Of course, even these self-representations as teachers or doctors invoke an image of expert professionals who have access to "discourses that command the power of truth" (Escobar 1995, 84). Yet the emphasis remains on informing farmers' decisions, based on the view that farmers need to be taught what is best for them.

However, there is a tension in the discourse of officials on this point, as local agriculture officers also speak frequently of forcing farmers to change their practices, and describe institutional practices employing "compulsion" *(kaTTayam)* to adopt new technologies. One of the most widespread of these practices is the tying of new farm inputs to sales of other things that farmers need. For example, many farmers have come to depend on the department for high-yielding varieties of paddy seeds. Unpopular inputs are tied to seed purchases, so that farmers cannot buy a kilo of seeds without buying five packets of weedicide, or biofertilizer, or whatever is currently being promoted. Older employees say this has long been the practice of the department, and that even the high-yielding varieties of seeds were once "thrust" on farmers in this way.

An oft-cited piece of lore among current and former extension workers concerns early efforts to introduce chemical fertilizers in rice production.

Agriculturalists strongly resisted applying chemicals, as they were believed to be harmful to the soil, causing it to be "ruined" *(keTu)*. In the face of this resistance, extension workers decided to bypass the difficult route through human consciousness. Abandoning efforts to convince farmers of the chemicals' advantages, teach methods of application, and arrange for them to obtain fertilizers, these workers went secretly into farmers' fields and applied the fertilizers themselves. As one former official described this process:

> Farmers had the opinion that if you put fertilizer, the soil would deteriorate or the crop would die immediately. So we would secretly put fertilizer in one corner of the field. Eventually we would ask the farmer why some plants in his field were taller and darker green. He would reply that there was more farmyard manure in those spots. Then we would reveal that we put fertilizer without his knowledge. This was our marketing business. Now it is the biggest business in the world.

Although the act of adding fertilizers without consent is rationalized here as an effort to teach or explain, the recourse to this practice also suggests an element of compulsion inherent in the marketing of new perceptions of nature and of need.

LOCAL OFFICIALS AS TARGETS OF DEVELOPMENT

Although local officials cast themselves as agents of development in relation to farmers, the officials themselves are targets of modernizing forces within the bureaucracy. In addition to the discourse of the traditional farmer, higher officials have a parallel discourse about the need to discipline and mold lower-level workers. Often these efforts to discipline are prompted by characteristics that local workers are seen to share with farmers. For example, local workers are described as insufficiently systematic and efficient in their work habits, and especially in their management of time and maintenance of official records. Higher officials also question extension workers' diligence in distributing new technological inputs, and their adherence to department guidelines and "norms." Perhaps most important, higher officials express concern that because village-level workers are not agriculture graduates, they do not have an adequate understanding of scientific agronomy and are ineffective in imparting accurate information regarding new technologies to farmers. These perceived deficiencies are addressed within the department through a range of institutional practices including setting of schedules, preparation of diaries and reports, administrative meetings, and training sessions. These mechanisms—involving "detailed processes of spatial organization, temporal arrangement, functional specification, and supervision and surveillance" (T. Mitchell 1991, 95)—attempt to mark out the terrain of the department, to ensure that all levels of the agricultural bureaucracy are pursuing coherent aims and generate effectiveness in fulfilling those aims.

As with farmers, efforts to transform the work practices of lower-level agriculture workers focus on the production of ingrained habits, understandings, and orientations. One higher official described a two-month USAID training program he had attended in 1969, which emphasized the inculcation of a "work ethic" and the idea that "you are being paid for work done." His aim was to move away from strict supervision and to pass on to his staff an internalized sense of responsibility, buttressed by the provision of external resources. With adequate resources—particularly vehicles—he felt that his staff would no longer have an excuse for not getting to meetings on time or fulfilling their schedule of village visits. Of course, these strategies to rationalize and manage the efficiency of workers rely on the same combination of transformations in consciousness and the provision of technologies—in this case, vehicles—that are employed to increase the efficiency of farmers.

Yet again, to the extent that such efforts are incompletely or ineffectively internalized, they are supplemented by more forceful mechanisms. In the speech of higher-level officers, there are frequent references to the need to "control" and "compel" lower-level extensionists to perform the tasks set for them. Extension workers, in turn, express resentment at what they see as overly strict supervision. For example, extension officers are expected to maintain diaries that note the villages they have visited, the farmers met, and the topics discussed. During my stay in the district, the department also introduced a rule that they should maintain a register of attendance at each of their village contact points. Although the ostensible purpose of the register was to provide a means for farmers to write their questions and requests, village-level workers felt that the real purpose was to keep them under surveillance. They were refusing to keep the registers, saying, "We are not under arrest."

Targets are the primary mechanism by which the development goals of the Department of Agriculture are enforced both on farmers and on its own employees. For every policy set by the state—for example, increased rice yields—physical inputs are distributed to farmers, often at subsidized rates. Each district is given a target of the number of inputs it must sell, and these are disaggregated by *taluk,* block, and village. It is ultimately the responsibility of the village-level workers to see that the targets are met. Unmet targets are compensated for by actual or threatened deductions from their paychecks.

The state Department of Agriculture claims that programs and targets are set in consultation with the districts, but district officials complain that they do not have much real voice in determining the content or quantity of targets and subsidies. State-level officials also assert that targets can never be too high because the department supplies only a small percentage of the estimated demand for various inputs. If anything, they say, the farmers are clamoring for more resources. However, as farmers also do not have a voice in decisions regarding the practices and inputs to be promoted, the resources the department provides are often not those for which farmers are clamoring. As a result, village-level workers find that the pressure to fulfill targets dominates their work. For example, a major headache for extension workers during my stay arose from farmers' pointed lack of interest in the newly introduced biofertilizers, a problem compounded by the fact that they were often released close to their expiration date. One village worker complained:

> We end up with three hundred packets on our hands and must pay the cost from our own pockets. We cannot say anything at feedback meetings because we cannot speak against the higher officials. The JD has ordered three thousand packets. Our superiors want to fulfill their targets, because they came from their superiors, so we can't ask them to change or cancel.

And another official added, "The higher-ups are not interested in the local situation. It is 'Do or die.' It is only to fulfill targets that we are being paid."

While the ostensible purpose of distributing targeted inputs and subsidies is to transform farming practices, they are at least as useful as a means of monitoring extension workers and holding them accountable. Targets were initially discontinued with the introduction of the Training and Visit (T&V) system, as the World Bank recommended privatization of the trade in agricultural inputs.[6] However, targets quickly crept back in, as they are seen by state and district-level officials as crucial to the conduct of business. "Targets have been used since the inception of the department," stated one official. "We must have some way of kindling the farmers to adopt new technologies. We also need a way of controlling the AAOs. Otherwise, they will be vagabonds. They won't care if a technology is introduced."

The transfer of technical knowledge is another day-to-day department activity that is permeated by simultaneous attempts to exert direction over farmers and local workers. According to the T&V system, the transfer of technical knowledge from agricultural research stations to farmers is to take the highest priority among the department's activities. As originally conceived, this flow of knowledge was to be two-way. In addition to bringing new research to the attention of farmers, extension workers were to give researchers feedback regarding farmers' needs. In this formulation, farmers do not possess "indigenous knowledge" as it has emerged in the development discourse of the 1990s (see Brokensha, Warren, and Werner 1980; Warren, Slikkerveer, and Brokensha 1995),[7] but the existence of "needs" provides a rationale for taking the perspectives of farmers into account and institutionalizing a channel for feedback.

Current department practice, however, diverges from this plan in several significant ways. First, the more difficult task of technological training of both farmers and lower-level officials is displaced by a concern with meeting targets. Second, when technological training is offered, it takes the form of information regarding recommended practices ("what to do") rather than a conceptual or explanatory system ("why to do"). Finally, the flow of this information between the department and agricultural communities is not two-way, as envisioned; rather, recommendations are simply handed down through the structure of command. These divergences, I argue, arise from two related causes: the invisibility of farmers' knowledge, and the struggle of the department to maintain coherence in its goals through to the lower levels of the bureaucracy.

Agriculture officials in Tamil Nadu refer to farmers' knowledge in many contexts, including meetings, extension visits, reports, and graduate theses.

However, both implicitly and explicitly, "farmers' knowledge" is defined as their awareness of department-recommended practices. The assumption that knowledge can come only from the state precludes the recognition that farmers may have understandings of their environment and natural processes beyond what they have absorbed from extension "messages." Among extension workers, this lack of recognition takes the form of an insistence that farmers do not have knowledge, but only practices. In their view, farmers are interested not in understanding why a technology works but only in seeing its effect on crop yields. The idea that—for farmers—"seeing is believing" leads AAOs to rely primarily on visual demonstration in their extension efforts.[8]

The invisibility of knowledge not sanctioned by the state also affects village workers, however, and prompts the extension of the development mission into the bureaucracy. In its design, the T&V system called for a significantly expanded and hierarchical bureaucratic apparatus, and an extension staff characterized by high levels of "motivation, energy, and abilities" (Cernea 1979, 231). When the program was implemented in Tamil Nadu, AAOs were supposed to have been agriculture graduates. Instead, I was told, members of the existing field staff were promoted into the system through lobbying on the part of their professional association. As a result, the task of extending technical knowledge to farmers has been marked by anxiety over the extent of AAOs' scientific training and the capacity of the department to ensure a minimum adherence to its extension messages.

Anxiety about the technological capacity of AAOs is illustrated most clearly in the context of department meetings. Almost all the frequent "training" and administrative meetings of the department are devoted to making sure that workers have achieved their targets. When they are not taken over by review of targets, training sessions focus on repetition of extension messages regarding appropriate types of fertilizers and pesticides, the amount, time, and method of their application, and so on. The typical format for monthly district and taluk-level training meetings is for the ranking official to pass out the recommended practices for a particular crop and read through them. The recitation is notably lacking in discussion—either to provide explanations of the processes by which the recommended inputs achieve their effects, or to solicit feedback from the officers being trained regarding the suitability or lack thereof of the practices in their areas. Questions are fired at extension workers to test whether they have absorbed the information, and wrong answers provoke reprimands. Department-sponsored training sessions for farmers follow a similar format, though with less of an overt disciplinary thrust.

This focus on practices cannot simply be attributed to differences in pedagogical style or the lack of an intention on the part of state agricultural extension agencies to impart a broader explanatory frame. All-India Radio morning and evening broadcasts for farmers, for example, promote the authority of scientific explanations of processes such as nitrogen fixing and pesticide absorption. And senior officials on visits to village training programs sometimes ask farmers questions about biochemical processes ("Where does nitrogen come from? How does it help plants grow?") and express concern that farmers understand why their recommendations are effective. However, training is displaced by targets, and knowledge is displaced by practices, because of concerns regarding control of village-level workers. As transfers of knowledge are relatively intangible, they are taken over by the measurable and trackable process of transferring financial and physical targets. Similarly, in the absence of confidence in AAOs' grasp of scientific agronomy, practices are seen as being easier to enforce and more susceptible to verification. Information becomes another means of disciplining AAOs, who—like farmers—are characterized by "wrong answers" that make them targets of the development process. The concern with impressing information on extensionists, in turn, reinforces the invisibility of farmers' knowledge, as the extent of farmers' awareness of recommended practices is the only type of information that has a channel for expression at department meetings and in official reports.

MEDIATING THE "STATE" AND THE "SOCIETY"

The institutional practices that seek to enforce coherence among levels of the department also serve to create the experience of a boundary between local officials and the farmers they seek to transform. As bureaucratic disciplines become increasingly finely regulated, the activities and orientations of local officials are differentiated from those of farmers, the content of these activities and orientations conforms more closely with the development vision formulated at higher levels of the state, and interactions between local officials and farmers become more restricted and inflexible. Building on the distinction between those with and without nahariham, and those who are agents as opposed to those who are targets of development, the institutional practices of state building progressively create a perception of a separation between the state and the society on which it must act. As Timothy Mitchell (1991, 95) argues, "The state should be addressed as an effect of detailed processes . . . which create the appearance of a world fundamentally divided into state and society." This appearance has real consequences, as both officials and agriculturalists perceive and act on it. But because it is ultimately the job of AAOs to induce farmers to act, many also end up "crossing over" the perceived boundary to make new technologies and forms of knowledge appropriate and comprehensible to farmers. In doing so, they draw consciously and unconsciously on the social ties and understandings they share with farmers.

Among older AAOs, who have experienced several phases of bureaucratic reorganization of the department, there is a pervasive view that relations with farmers have worsened since the officers' early days of government service. With the elaboration of the agency hierarchy, larger numbers of staff, more comprehensive extension coverage, and an intensified regulation of the rhythms and results of their work, relations with farmers have become ever more constrained. Increases in targets, and the displacement of the officers' role of extending knowledge by one of promoting new inputs, are identified as creating particular tensions between local officials and farmers. Extension workers feel that pushing inputs and subsidies has weakened their credibility as a source of technological expertise, and that farmers have come to see them as "salesmen" for the latest government goods. On the other hand, they fear that if they gave up the practice now, farmers would not even approach them. As one AAO remarked:

> Our most important job is to teach technologies to farmers. The second most important is giving subsidies with the help of the World Bank, and the central and state governments. But when farmers see me, they do not ask about technologies. They say, "Do you have any new seeds, or fertilizers?" If they are available, they ask, "Is there a subsidy rate?" If I say no, they go away.

AAOs contrast this state of affairs to their earlier experience, when their schedules, movements, technical recommendations, and general interactions with farmers were less strictly regimented and monitored by the department. Then, I was told, extension officers "moved with farmers as one among them. Now the number of posts have increased, but the farmers no longer help us."

With the progressive elaboration of the institutional practices of the state, local officials no longer "move with farmers as one among them" but must negotiate two worlds—those of "state" and "society." One indication of the divergence of these two social orders may be found in the contrasting strategies of self-positioning that local officials adopt in the course of their work. Some individuals remark that they approach farmers as representatives of the government in their efforts to convince them to purchase inputs or adopt new practices. "It is better for the communication of the message if farmers see me as an Agricultural Officer," one explained. "They will have respect for me and look up to me as an official person who knows some things." Yet others find that they are more effective in their work if they position themselves as fellow farmers and draw on the strengths of their local connections and relationships. Thus one AAO commented that when he was given a target of thirty acres of line planting by his superior, the farmers in his area complied out of friendship for him. Another said that farmers will purchase unneeded inputs "for my face." Many officials also assert that farmers are more likely to adopt a practice if they present themselves as farmers who have used it in their own fields rather than as officials who have seen it demonstrated at an agricultural research station. Thus, positioning as an official of the state carries with it the prestige and authority associated with state-sanctioned knowledge, whereas positioning as a farmer calls on shared experiences, mutual obligations of friendship and affinity, and relevance of the technologies to local conditions.

Even as the disjuncture between state and society grows, the need remains for state actors to engage with the values, understandings, and motivations of agriculturalists to achieve their development goals. Indeed, intensified control of local officials only serves to increase the pressure on them to engage with farmers, because they are more strictly monitored in the fulfillment of their assigned tasks. As a result, there is a considerable amount of slippage of department norms, regulations, terms, and strategies as AAOs seek to render their programs intelligible and attractive to farmers. Sometimes, when officials are not able to elicit the desired response from farmers, this slippage simply takes the form of false or modified reports. For example, AAOs are required to hold evening meetings in each village, but as farmers do not attend, the officers simply mark in their diaries that a meeting was held. "If I meet a farmer at the tea stall and exchange a few words," one AAO joked, "I write 'meeting' in my report." Higher-level officials allude to the keeping of "soft records" in order to show compliance with input distribution demands in the face of a lack of interest by farmers.

Other means of negotiating the lack of fit between the worlds of office and village include bending program rules, altering program strategies, and shifting terms of understanding. In addition to the practice of input "thrusting" I have described, workers sell targeted items to people who are not eligible—such as large farmers—or people who have already received their quota. Extension workers also trade tips among themselves about how to divert targeted inputs to other uses. In one case, an officer who was having trouble selling his allotment of "improved" wooden plows was told by his colleagues to point out to farmers that the subsidized price of the plow was less than the price of an equivalent amount of firewood (implying they might be used for the latter purpose). Village-level workers are constantly being reprimanded for these types of infractions, but they argue, "We have to meet our targets. In our office, it is not possible to maintain standards *(niyayam)*."

Although the training of AAOs does not emphasize explanation, local officials also find that they must engage with farmers' conceptualizations of agronomic processes to convince them to adopt new technologies. In doing so, local officials negotiate the disjunctures between the idioms of scientific agronomy promulgated by the department and the conceptual frameworks that guide farmers' agricultural practices. This mediation, I argue, takes the form of a subtle and contextualized shifting between explanatory modes rather than a "translation" between systems of knowledge that local officials perceive as bounded and discontinuous.[9] Indeed, the mediation may be all the more effective because it is not structured or explicit.

Formal acts of translation are undertaken by various branches of the agricultural extension service in their efforts to present information in ways that are accessible to farmers. All-India Radio, for example, uses Tamil terms for many introduced inputs and practices, and it broadcasts the multiple local names of pests and diseases—which vary within the district—along with the terms standardized by department usage. According to AIR officials, their broadcasts have tried increasingly to engage with local communicative contexts and performance genres, which are used to promote both specific technical practices such as Integrated Pest Management and broader government policies.[10] Extension trainings have also tried to introduce measures that are grounded in everyday objects rather than abstract standards. For example, they teach farmers that a matchbox equals ten grams, and that one *san* (a measure of distance based on the span from outstretched thumb to small finger) is fifteen centimeters long, or recommend that workers place a cycle tire in the fields and plant twenty-two seedlings inside it to achieve the optimal spacing of sixty-six plants per square meter.

The acts of "translation" undertaken by village officials, however, are of a less formal sort. Because the concepts underlying farmers' practices are shared, or at least understood, many local officials move easily between the "scientific" idioms in which department recommendations are couched and local explanatory frames. This ease is enhanced by the contexts of interaction of farmers and AAOs. Typical interactions take the form of casual conversations in village homes, teahouses, and fields, in which both extension workers and farmers orient their speech and understandings to the other.[11] Farmers may adopt standardized measures and biochemical terms with officials that they do not typically use with one another (though the degree of familiarity with these terms varies widely). More often, it is AAOs who—trying to convince or explain—shift to agronomic understandings, terms, and measures that are characteristic of farmers' speech.

One of the most pervasive underlying frameworks of local agronomic understanding consists of the use of humoral classifications. Knowledge of rice cultivation is organized according to principles of humoral agronomy in which qualities of heat and cold, and wetness and dryness, are balanced and managed. Most AAOs participate in these understandings and will describe cultivation processes and both organic and inorganic inputs in terms of their heating or cooling effects on plants. Another characteristic mode of explanation among farmers involves the use of body analogies. Thus extension workers compare the application of complex fertilizers in the seedling nursery to the feeding of nutritious foods to a pregnant woman, or the split application of fertilizers to the need for people to eat three separate meals, and not all the day's food at one time. With regard to terms and measures, many AAOs start a conversation by using English terms but then switch to the terms, or measures of distance, volume, area, and yield, used by farmers themselves. As one extension worker explained: "When we tell them to put kilos, they ask, 'How do we know kilos?' So we tell them how many *pakka* [a local volume measure]. Even in my own house, we are using pakka."

As a result of the mediating role played by extension officials, neither they nor farmers operate with a strong sense of "science" as a body of knowledge that is distinct from "indigenous knowledge." The perception of difference is further minimized by the fact that officials, by and large, do not recognize farmers' knowledge, and by the lack of an emphasis on "scientific" knowledge in the training of extension workers and farmers. AAOs acknowledge that farmers do not share the models and terms of bio-chemical agronomy, but this problem is not foregrounded because they see their development mission as one of modifying practices rather than transforming meaning. Thus much of the discussion regarding agricultural technologies between AAOs and farmers can take place at the level of practice. At times, farmers and officials in the same conversation offer widely divergent explanations of an observable agricultural phenomenon such as a crop disease. However, if the divergences in understanding do not entail different practical interventions, they tend to be ignored. One significant result of the focus on practices is that technological change can occur without a corresponding change in agronomic understandings. Chemical fertilizers, for example, have become assimilated into humoral understandings as an extremely heating input and are managed accordingly.

Although the institutional practices of the Department of Agriculture thus do not contribute to the emergence of a boundary between "scientific" and "local" agronomic knowledge, the boundary that marks state-sanctioned knowledge from that of the village is reinforced. Despite their acts of mediation, officials at all levels ultimately assert the authority of the explanations and terminologies promulgated by their department. For example, district officials at a village training event expressed concern that farmers were not answering "correctly" when asked for their explanations of various agronomic processes. For these officials, there was a correct answer, and its repetition was proof that they were accomplishing their work satisfactorily. With regard to local terms, AAOs admitted using them but insisted that farmers eventually must learn those of the department because—as one put it—they are "universal." Another added, "They should change to the official language. It is not possible for us to change to their terms. We cannot violate the government-given name." Many villagers have also come to recognize the authority of state-sanctioned knowledge and its association with progress and modernity. Differentiating himself, in turn, from those who are less developed, one farmer remarked, "If we go to town and people are talking, only with nahariham can we follow their speech. In some small villages, the people have no nahariham and they will not understand any message given by the government or on the radio."

CONCLUSION

An examination of efforts to transform the agrarian environment of a district Department of Agriculture in Tamil Nadu reveals that the state cannot simply be assumed to be a monolithic agent, clearly distinguished from the society it seeks to influence. A focus on the position of local agricultural officials, in particular, calls into question the possibility of viewing the state as a coherent whole and marking a neat boundary between "state" and "society." Local officials both differ in their understandings and motivations from officials at higher levels of the state and share significant commonalities with farmers. With regard to the role of the state in promoting development, local officials are simultaneously agents and targets. They are agents of development in that they affiliate themselves with the aims and ideologies of development and are on the "front lines" of efforts to transform local agricultural practices. At the same time, they share with agriculturalists social, political, and conceptual commonalities that make the officials targets of parallel efforts toward increased efficiency and productivity on the part of higher officials.

Efforts to increase the productivity of both farmers and local officials involve not simply the production of material goods but also the production of deeply ingrained orientations, identities, and aspirations. To the extent possible, these transformations occur through a slow process of internalization. However, because of the pressure to "develop" (to "catch up," to "take off"), these gradual methods, I argue, are supplemented by more active mechanisms of control. Often the same mechanisms—such as targets—operate on both farmers and local officials. However, these mechanisms are more successful in constraining and shaping the actions of AAOs than of farmers because the former are more susceptible to the state's practices of control.

Through these everyday institutional practices, the boundary between state and society becomes less blurred. That is, the daily routines, rhetoric, and goals of the Department of Agriculture increasingly differentiate the social world of extension officials from that of farmers. However, as local officials are charged with inducing farmers to act, they must take a variety of steps to mediate the lack of fit between the perspectives of farmers and development visions formulated at higher levels of the state. These steps include circumventing department guidelines and rendering programs and practices in terms that are comprehensible and attractive to farmers.

These efforts to negotiate disjunctures, however, are almost always read by state actors—even those who perform them—as "corruption" rather than as "innovation." Rather than seeing the intermediary position of local officials as an opportunity to modify development efforts so that they are more appropriate to the needs and understandings of farmers, the discourses and practices of the district agricultural extension service serve to attenuate connections with rural actors and bolster its own unity and cohesion. An analysis of the institutional mechanisms by which this separation and cohesion are asserted helps to address the question of how the state can be disaggregated yet retain its power to achieve ordered effects. What is lost in this process is the possibility that development goals could be informed by local values and understandings. Under these circumstances, the mediating role of local officials serves mainly to bring the ideologies of development closer to the experience and aspirations of agriculturalists.

NOTES

1 Colin Gordon, in his analysis of Foucault's writings on power and knowledge, identifies these two problems as "an illusion of 'realization' whereby it is supposed that programmes elaborated in certain discourses are integrally transposed to the domain of actual practices and techniques, and an illusion of 'effectivity' whereby certain technical methods of social domination are taken as being actually implemented and enforced upon the social body as a whole" (C. Gordon 1980, 246). In accepting these "illusions," critiques of development risk sharing the assumptions of state control and governability that underlie development efforts premised on state intervention.

2 See, for example, J. Comaroff 1985; Comaroff and Comaroff 1991; Taussig 1980; and Ludden 1985.

3 See Gupta 1995 and T. Mitchell 1991, however, for groundbreaking analyses of the everyday practices of the state.

4 Research in India was conducted from July 1993 through December 1995 and was supported by a Fulbright-Hays Doctoral Dissertation Research Abroad fellowship and a Wenner-Gren Foundation predoctoral grant.

5 I use the terms *extension worker, village-level official, local official,* and AAO interchangeably in this study to describe them.

6 The World Bank privatization policy assumes the existence of a consumer orientation among farmers and their responsiveness to the technologies promoted by the state. The department's methods, however, implicitly acknowledge that the benefits of development might not be readily embraced by farmers.

7 See Agrawal 1995 for a critique of the conceptual weaknesses of efforts to define a monolithic "indigenous" knowledge and to separate it from a similarly monolithic body of "Western" science.

8 Extension workers themselves lament that this bias toward "seeing" has oriented farmers toward invasive technologies with relatively dramatic visual impacts.

9 Although a consideration of local agronomic knowledge is outside the scope of this paper, I note that its invisibility arises in part because this knowledge is not, as Paul Richards (1993) puts it, a "satisfyingly complete, freestanding, 'agricultural knowledge system.'" At the same time, agronomic practice is guided by durable orientations and broad frameworks of understanding (see Bourdieu 1977).

10 The most striking, perhaps, was a paean to the General Agreement on Tariffs and Trade (GATT) performed in the style of a bow song.

11 As Voloshinov (1973, 86) states, "A word is a bridge thrown between myself and another. . . . A word is territory shared by both addressor and addressee."

Darren C. Zook

Famine in the Landscape: Imagining Hunger in South Asian History, 1860–1990

Much of contemporary writing about famine and the environment is embedded in texts and narratives that are often associated with the struggle for social or political justice. The environment must be sheltered, and famine victims saved, and those who are responsible for causing environmental degradation or perpetrating famine in the present or in the past must be exposed and somehow brought to justice. But the master narrative of justice that these texts collectively produce is a much more difficult and jagged rhetorical geography than many studies of the environment or famine would have us believe. Across the vast landscape of history, there is no neutral terrain, no highest mountain, from which to observe and dispense justice. At best, there can be only a different mapping of the same landscape that has been mapped so many times before. This article looks at famine in relation to the indeterminate realm known as the "environment"—specifically the agrarian environment—and argues that in relation to the textual maps of famine in the environment, there exists no permanent standard for judging which maps are more accurate or better than others. The master narratives and cartographies of famine are therefore not bedrocks in the natural foundation of some primordial struggle for justice but morally normative descriptions that can be read as lamentations for lost (imagined) communities shrouded in the mists of inchoate, disremembered pasts.[1]

Specifically, I focus on the political and historical process of constructing a new moral environment in the South Asian agrarian landscape, and I argue that the famine narratives that were produced in the latter half of the nineteenth century were descriptions not so much of the particular famines as specific events but rather of the political agendas of various groups of witnesses. The clash of these political agendas ultimately did very little to end the problem of famines but did manage, perhaps unintentionally, to produce a new image of India first as a "land of famine" and later as a "poverty-ridden" nation. At the same time, these narratives serve as guidebooks to charting the changing moral landscape within South Asia itself, and through which South Asia would be judged by others.

In the conclusion, I address the question of how to reconcile the desire to believe in a more moral past (itself a product of a more immediate search for justice) with the proclivity toward disbelief and mistrust—of narratives, testimonies, and "texts" in general—that postmodern historiography has proposed. In some sense, this is yet another facet of the struggle between faith and reason. The answer, I suggest, is neither to throw up our hands and bewail our modern laments, nor to construct imagined pasts through moral polemics about more wholesome but now lost worlds, but rather to develop an approach to famine that contains no neat endings, no clear villains, and no pure heroes, and yet does not occlude the possibility of justice, if that is what we seek. In other words, we must accept the possibility of writing narratives of justice that lack a sense of closure, but in which suffering and justice coexist.

FAMINE IN THE HISTORICAL LANDSCAPE

How shall we describe the hunger which the poor experienced? What they felt was rarely transcribed by themselves, and those who took up the pen described for the most part only what they saw and heard. (Abel 1974, 321)

This comment from Wilhelm Abel describes succinctly the difficulties that anyone who approaches the study of famine and hunger must face. Those directly affected by calamitous hunger or famine scarcely possess the strength or leisure to put down in writing what they are suffering, and the witnesses who do possess this strength and leisure, regardless of their motives, are always and necessarily outsiders who describe what they see through a lens of many imperfections. Moreover, in cases where we possess a large enough corpus of writings from different perspectives about a particular famine that might help us to "correct" our imperfect lenses, a situation that begins to occur with increasing frequency beginning in the nineteenth century, there is always the risk of producing an emotional numbness, a fatigue of sympathy, through a seemingly endless stream of morbidity accounts and descriptions of distress. The more we obey the historical compulsion to record the minutiae of the calamities that befall humankind, and the more widely these accounts are circulated, the more we run the risk of evoking not empathy but complacency and inaction.

To resist the drift into banality in the midst of catastrophe that the similarities of famine narratives represent, we need to address the question of how famines differ from one another. The most obvious answers—the degree of mortality, geographic scope, chronological duration, intensity, or severity—unfortunately do not take us very far because they elicit comparisons between common and quantifiable indicators; this tells us not how famines differ but how they are more or less the same. Such comparisons end up trivializing the experience of famine by leading to questions such as whether a famine that claims 500,000 victims is not as "bad" as one that claims 600,000. To avoid this line of reasoning, I suggest we shift our understanding of famines slightly to take into account the differing ways in which specific famines are *represented*. To be sure, the switch to an understanding of how famines are represented runs the risk of transforming the calamity of famine into a sterile academic exercise in the structuring of "regimes of representation," whereby the representation of an event takes precedence over the event itself. My point here is that famines are represented differently because of the different moral environments within which they take place and through which they are described and recorded. Indeed, the belief that famine can be prevented or even eradicated is itself contingent on a significant shift in the moral landscape of famine in relation to the agrarian environment.

One of the most important forms of famine depiction is the eyewitness narrative account, a genre that emerged in the nineteenth century largely on the back of popular mass media style journalism. Yet historically the use of mass media in representing famines has had a bittersweet aura about it. Some famines inevitably become "more important" (or more sensational) than others, for instance, in the case of Ireland, Finland, and highland Scotland: although all three suffered major subsistence crises in the mid–

nineteenth century, only Ireland's is universally recognized as historically "significant" (Ó Gráda 1992; Devine 1988). Images of a specific tragedy become emblematic of the society or culture of a region or nation, as in the case of Bangladesh, Ethiopia, and Ireland (Morash 1995; Sorenson 1993; Greenough 1982). Local narratives and representations become suspect or drastically (if unintentionally) altered in their translation into mass channels of communication. During the second half of the nineteenth century, for instance, when British journalists such as William Digby and F. S. Merewether traveled to India to see famines firsthand, and Indian luminaries such as Romesh Chandra Dutt and Dadabhai Naoroji traveled to England to present the sufferings of India's poor and hungry to the English public, the messages contained within even the best-intentioned or well-crafted famine narratives became quickly and hopelessly confused. As the second part of this chapter will show, although mass media provided a powerful tool in the hands of both proponents and opponents of British famine policy in India, the final image of India that emerged from the polemical battles in the pages of the press, and the final representation of India's historical and social landscape both in print and in graphic image, was that of a poor and starving country unable to fend for itself. Strangely, this was an image that neither side originally set out to prove.

Local representations of famine, along with other "folk" type sources such as oral histories or vernacular songs, are often assumed to be more truthful or accurate in their depictions of famine causality or suffering. Yet the difference is not actually one of "the truth" versus a nonlocal fabrication but rather one of two separate narratives, perhaps equally fabricated, but serving different purposes and fulfilling different moral and political agendas. In other words, local representations of famine turn up different interpretations that do not so much resist or disprove mass representations as confuse and reconstrue them. If we look at various representations of the Bengal famine of 1942 to 1943, for instance, we can see how the attempt to create a coherent, seamless narrative ultimately reveals itself to be an awkward, imperfect process. One recent volume on the Bengali famine that attempts to place the calamity squarely in collective memory by bringing together various modes of recollection and representation begins its section of "living memory" by invoking the name of the "great calamity in our golden land" and then offering to "correct the muteness of history" by assembling the "fear of scarcity and the palpable pathos of (mass)" death into the body of history and remembrance."[2] But whose memory and whose history are being assembled? In spite of efforts to situate the Bengali famine into a narrative of anti-British nationalism by placing blame for the calamity on the indifference and cruelty of foreign rule, the Bengal famine remains in the end a Bengali experience, a local history that does not quite "fit" into the broader narrative of Indian nationalism. Indeed, one particularly popular Hindi-language poem designed to bring the suffering of Bengal into national (Hindi?) memory cannot seem to escape the uniquely Bengali aspects of the experience: "This is Bengal— / look upon it with enraptured eyes / with quivering throat / with ecstatic voice / here is the nation-song sung by the poet— / *vande mataram.*"[3] Here, the "nation-song" remains ambivalent. The phrase *vande mataram* (hail to the motherland!) was taken by Rabindranath Tagore from Bankimchandra Chatterji's nationalist novel of Bengal, *Anandamath,* and made into India's first national anthem; later, however, it fell from favor as a national anthem because it was too "local" and reminiscent of the (violent) Bengali experience of nationalism. Here, different languages, different vocabularies of memory, and different experiences of nationalism all collude as much as they divide, and no consistent, stable narrative can emerge from such unstable conceptual territory.

Even the narrative of blame in the Bengal famine is not as straightforward as it appears. Typically, the argument goes, British adherence to the exigencies of economy and trade allowed grain to flow out of Bengal (or out of famine districts within Bengal) when it was most needed, and the lack of timely relief allowed the most vulnerable populations to suffer and perish quickly throughout Bengal. Yet a report published just after the conclusion of the famine cites internal causes as much as (if not more than) it cites the lack of state intervention. Specifically, black marketeers and profiteers within Bengali society, acting "anti-socially" and "unpatriotically," are faulted for failing to heed the civic call of (Bengali) nationalism and for causing the suffering of their fellow Bengalis.[4] In this context, narratives of blame that focus exclusively on British imperialism may be designed more to avert attention away from the faults of Bengali society than to offer a dispassionate account of famine causality.

From a different angle, that of assigning praise (rather than blame), and in a different setting, that of the Sudan during the mid-1980s (a period often associated with the intensification of famines in Africa and the rise of a number of private charity efforts in the West such as Live Aid and other mass media crusades), Alex De Waal notes:

> Before 1985, Darfur had never known a major food aid programme. The arrival of food aid from outside, known as *Reagan* after the man who had supposedly donated it, was greeted with bafflement and surprise—and delight. The idea of the government (which, to the people of Darfur, embraces the "international community" as well) having an obligation to supply villagers with food was a novel idea. "Who is this Reagan?" said one farmer. "He ought to be promoted!" (De Waal 1989, 204)

As will become clear in the following discussion, it would be quite wrong to interpret this statement as implying that before the 1980s Darfur had not experienced major famine simply because it had not experienced a major food aid program. Similarly, it would be facile to interpret the anonymous farmer's praise of Reagan as an unfortunate occurrence of "false consciousness" or the power of the representation of American (media) hegemony; we may have to read it just as it is—praise of someone who is seen, rightly or wrongly, as providing charity and relief during calamity. Alternatively, if the commemoration of Reagan in local oral accounts of the famine is somehow a misrepresentation, then how can we assume that oral accounts of other famines that praise local leaders are somehow more accurate and authentic?[5]

Problems of interpretation and verification occur not just between different types of situational perspectives (local versus nonlocal) but also between different types of historical or cultural perspectives. The comparison of different records of famines over different historical periods or in different cultural contexts is so fraught with difficulties that it threatens to render the attempt to discern responsibility for famines nearly impossible. One cannot compare, for instance, the absence of complaints or solicitations for relief rice in Mughal India with their replete presence in colonial records and assume that this means that Mughal kings took care of the populations during famines and British colonial officials did not. It *might* mean this, although the historical evidence is certainly weighted against such a conclusion. But it is also possible, and perhaps probable, that the absence of solicitations in Mughal documents means only that no one expected the Mughal officials to supply any relief, so that it was pointless to ask, or that court scribes whose job it was to record the "event" portrayed only what would make Mughal rulers or patrons look magnanimous and charitable.[6] Ancient texts are often enlisted in the struggle to develop comprehensive historical arguments about who or what is responsible for the occurrence of famines in the modern era. In India, Sanskrit texts such as Kautilya's *Arthasastra* or Tamil texts such as *Tirukkural* or *Purananuru* are often recruited to "prove" that famines did not exist in India's ancient past or, if they did exist, were largely rendered benign through "traditional" acts of social cooperation and civic acts of benevolence and redistribution.[7] Although Kautilya has certainly listed a set of principles to which a ruler should adhere during famines and other calamitous "acts of God," it is unclear whether any of these were actually followed in practice, or, in the case where we do have an example of them being followed, whether this represents what actually happened or what a court-appointed scribe was "encouraged" to record.

The proclivity to resort to the use of texts that are seen to be somehow outside the boundaries of hegemonic centers (such as colonialism) reflects a desire to find a moral arena from which to critique and deconstruct so-called hegemonic representations of famine. This desire is common to all critical narratives that possess as their motive the search for, or rendering of, some form of social or historical justice. Yet if those areas outside the bounds of hegemonic centers of gravity—the areas that are purportedly the source for local narratives such as oral histories, folk songs, and so forth—are no more truthful or accurate in their representations of events, then the moral center for critiques of hegemonic distortions of representations evanesces. Two of the most powerful and persuasive "master narratives" of social justice in relation to food security and famine prevention—the theory of the moral economy and the theory of exchange entitlements—are rooted within a series of assumptions that have not been rigorously proven as universally true or historically valid (E. Thompson 1971; Scott 1976; A. Sen 1981). By singling these out as master narratives, I do not mean to suggest that either of them is intrinsically invidious or insidious or that they produce the same kind of violence or distortions often associated with other master narratives such as colonialism or nationalism. What I do mean to suggest is that the issue of assigning blame, historically or otherwise, for famine causality and the task of procuring justice in the form of famine prevention may not be reconcilable projects. Famine prevention is not the righting of historic wrongs but a justice of the moment—even of the colonial moment (ironically). The narratives are too unstable—we really have no more reason to trust an oral history than we do to mistrust a colonial document. The conclusion this points to in relation to famine narratives is that we cannot use oral histories or histories from below to "deconstruct" the master narratives of colonialism, but rather we can only deconstruct them *in the same way* that we deconstruct colonial master narratives.

In short, there is no clear cultural or historical pattern in relation to environmental degradation or famine relief and prevention, and from among the myriad environmental or charitable values that different cultures and societies profess, none is absolute, and none is universal.

THE CASE OF SOUTH ASIA

Most of South Asia, like most of Africa, has been associated with the persistence of widespread poverty and hunger. Few would deny that hunger and poverty are in fact widespread in parts of south Asia; the debate centers mostly around why they are there and what their persistence represents. Some see them as legacies of colonial rule, and others see them as the product of "backward" agricultural practices and social customs. The general debate about the nature and causes of hunger and poverty in South Asia has its roots in the last half of the nineteenth century, when the issue of famine causality was linked with nascent nationalist polemics about the effect of British rule on India's economy. In effect, the two strands of rhetoric that emerged during this period—one that roughly represents the colonial side, and the other the nationalist—represent two competing master narratives emanating from different moral universes. Like the moral economy argument or the theory of entitlements, the master narratives concerning famines, hunger, and poverty in colonial India were designed to identify the "source" of current troubles and document the ways in which injustices have been perpetrated and the ways in which justice might be rendered. The colonial side was seeking to vindicate both colonial rule and economic liberalism; the nationalist side was seeking to blame colonial rule and to construct a precolonial mythology of economic prosperity and social justice. For the latter, famines, hunger, and poverty had to be the fault of the British, for otherwise they could only be self-inflicted, something that would threaten the nationalist agenda by acting as a rather indelible taint on its rhetoric about precolonial prosperity and cultural pedigree.

My argument here concerns the general transformation of the perception of south Asia in general and India in particular from a region visited by periodic famines to one afflicted with chronic poverty; the argument can be summarized as follows. During the last half of the nineteenth century, famine became inextricably connected with the landscape of India. Ironically, it was placed there as a result of the debate about who was responsible for the famines that occurred in India at that time. What began as a polemical conflict to pin responsibility for the famines onto British rule ended up creating an image of a starving population unable to fend for itself. Roughly after the turn of the century, as the four-decade-long run of famines came to a close, the geographic landscape of famine was transformed into the mental landscape of chronic poverty. In studying the ways in which competing images of India become established as legitimate or authentic representations, it is essential to include images of self-representation as well and to push beyond the tendency to focus only on how "the West" represents "the non-West."

Aside from the question of images and representations, the new landscapes carved out by famine and poverty in India became the setting for the expansion and intensification of the historical project of economic and specifically agrarian development. The logic here was that the poorer India was, and the more hungry the people, the more development India needed. To conquer India's poverty and to eradicate the famines that had afflicted the country since "time immemorial" would thus require nothing less than the conquest of nature. This conquest of nature would come first in the form of altering the landscape through massive rural development and public works projects and later, roughly at the start of the twentieth century, in the guise of social reform and community development projects.

Famine

In 1876, in the midst of the Great Famine of 1876 to 1878 in southern India, there appeared in India and England a poem in the form of a pamphlet that was designed to elicit both sympathy and money for "relieving the famine-stricken ryots of India." Significantly, it was an appeal addressed only to

Englishmen, as they alone were "able to afford it." The rest of India, from the rajas down to the *ryots,* were apparently considered too impoverished or too indifferent to make donations or provide relief, and hence the task of "saving India" had become the responsibility of the British. The significance of the poem lies less in its literary merits, which are few, than in its observations on India's famines. The poem begins thus:

> I sing the tragic, but pretend no art;
> For simple facts, unpolished, move the heart.
> A hardy swain—his features vexed and wan:
> Languid and lifeless eyes: his vigour gone:
> His limbs betray an attitude of cares:
> Hunger—a deadly canker—only tears
> His bowels starved: and pain inflames his soul:
> His mind, confounded, dimly sees its goal—
> This famished peasant oft attempts to stand,
> And anxious sees how sad the thirsty land![8]

This poem—and there are ten more pages of it—illustrates in brief some of the attitudes toward famine and poverty in India in the latter part of the nineteenth century, attitudes that, as we shall see, still persist in much of the literature on India. We see in this poem, for instance, the decrepit and helpless Indian peasant, meant to represent the rural population of India as a whole as it suffered through the famines of the late nineteenth century. We also see the depiction of famine as something that appears naturally in the "thirsty" landscape of India. The logic embedded in this line of argument is that if famine is a natural part of India's landscape, and if the Indian ryots, and perhaps all Indians in general, are poor on account of these famines, then India must be a nation that is naturally poor.

The Great Famine of 1876 to 1878 was a crisis moment, indeed a conjunctural moment, in the moral trajectory of south Asian agrarian history, primarily because it brought to the fore the issue of moral responsibility both for the causation of famines and for the provision of famine relief. The determination of historical responsibility for famines is an integral part of the process through which representations of India were transformed from a country with a sumptuous if decadent culture to a country with hunger and poverty in every nook and cranny of its landscape. The Bengal famine of 1770, for instance, did not become a political issue until nearly a century after it occurred, when it suddenly became a fictitious starting point for the record of famines that came to represent the effects of colonial rule in India for a growing chorus of nationalist critics. In 1770, it seemed oddly congruent with emerging "colonial" representations of India and largely nonexceptional in local representations and histories; in 1870, when debates about famines and poverty were part of public and political dialogue, the Bengal famine suddenly became a useful venue for arguing over the role of famines in Indian history and the prevalence of poverty in Indian culture.[9] It is not entirely coincidental that Bankimchandra Chatterji's novel *Anandamath* appeared during this period: not only does the novel take up the issue of responsibility for the 1770 Bengal famine, but also, and more significantly, it divides up this responsibility along ethnic, religious, and nationalist lines. Those who caused the famine were "outsiders" perpetrating criminal acts against innocent "insiders"—in this case Bengali Hindus—who by default became victims of criminal aggression, in turn opening up the possibility that the nationalist struggle (at least in Bengal) would elide the struggle for moral justice, a struggle for which the issue of famine responsibility served as a convenient, if somewhat contrived, catalyst.

Obviously, famines had occurred in India in both the recent and more distant past, but somehow the famines of the latter half of the nineteenth century were *perceived differently* and hence *represented differently*. For the British administration, the transformation in representation and perception can be dated from roughly the 1860s, when a series of famines spread throughout large portions of India and would continue to do so periodically until the turn of the century. To the official mind, these famines were seen as evidence of India's inherent susceptibility to natural disasters. The drive to draw up a comprehensive famine code, which would consume a large portion of administrative monies and energies for much of the last half of the nineteenth century, was designed to serve three simultaneous ends: it would shield India from the capriciousness of nature, it would breathe new life into the fledgling experimental program of agricultural development begun in the 1860s, and it would represent and codify the nascent belief that the paramount duty of any "civilized" government in India or elsewhere was "to save the lives of the people over whom they ruled."[10] Many writers of this period commented on the apparent "naturalness" of famine and want in the Indian environment. Robert Burton

Buckley, a retired chief engineer for the Public Works Department, looked back on his career in southern India—a career that spanned the last two decades of the nineteenth century—and noted the many areas of southern India that were "peculiarly liable to famine." Singled out in this group were Bellary and Anantapur—names that in the last decades of the nineteenth century became virtually synonymous with famine. Buckley, like so many others, advocated an expansion of agricultural development through ambitious irrigation schemes, noting that the "value of irrigation works in protecting particular tracts from famine has a political, administrative, and humanitarian value which cannot be gauged by money."[11] Echoing this sentiment, A. T. Mackenzie, in compiling a history of the Periyar irrigation scheme, noted that the historical records of the Madura district make "constant allusions to famine and scarcity" but through the great engineering projects of the colonial period, which brought water into the thirsty areas of the landscape, "a large number of useful human beings are *practically secure from want.*"[12] The wording in these quotations is both important and representative: the task of famine prevention was to protect "particular tracts" from the irregularities of nature, and to bypass nature with the use of science and engineering. A wealthy landscape was one in which water flowed freely and predictably, and hence famine, wherever and whenever it occurred, was in some sense a problem of landscape.

In the meantime, as the British were simultaneously blaming nature and taking credit for the "benevolent and charitable" acts of famine relief, most of which consisted of having famine victims build irrigation channels such as those advocated by Buckley and Mackenzie, other groups, especially those who would later be known as economic nationalists, were beginning (from roughly the 1880s) to blame the British administration for the occurrence and the virulence of these famines. Fictional pasts were quickly mobilized to show how ancient and indigenous rulers dispensed unlimited amounts of relief and charity and remitted all taxes for the duration of the famine. Romesh Chandra Dutt, to name the best known, compiled enormous volumes of statistical evidence to show how the oppressive nature of the British administration's taxation and how the continual drain of the wealth that rightfully should have stayed in India were the primary causes of the famines throughout India. Dutt's "evidence" here was the *hukumnamas* (village records) that supposedly represented actual historic and customary practice in the villages of the precolonial Indian countryside. These he counterpoised with his assessment of the experience of British rule, "under which a large, industrious and civilized population were rendered incapable of improving their condition." [13]

All these texts center around the issue of responsibility: for Dutt, the responsibility for famine was located in British land policy, whereas for Mackenzie and Buckley, it was located in the land itself. Yet all strangely agreed on one thing: Indians had somehow become or were seen to be incapable of improving themselves or preventing famine, and hence, while the debate on who or what caused the famines continued to rage, the debate over who should end them was over before it started—the task of eradicating famine in India was the responsibility of the British. The colonial state had become the ultimate patron of well-being, responsible for ensuring security from want and, later, enhancing the quality of life. Although in its cruder moments the argument for British responsibility has devolved into a charge of British genocide in India, in fact, even the most ardent polemicists of the time—and here I will single out Dadabhai Naoroji—used the issue of responsibility in such a way as to imply a new and enhanced legitimacy of colonial rule in India. In some ways, the most important shift recorded by Naoroji was not the shift from wealth into poverty but the shift from "despotic" to liberal principles of government in the representation of benevolent government in the Indian context. This in turn was buttressed by a new sensibility toward social obligations and charitable service, which were themselves part of the continually evolving agricultural development schemes. Naoroji, in his *Poverty and Un-British Rule in India,* did not counterpoise British rule and famine policy against an indigenous model; rather, he held the British up to their own ideals—ideals that he felt were good and necessary for India—and judged them to be out of sync, or, as the title of his work states, un-British. To be British, according to Naoroji, was to care for the people of India, to save them from famine, and to ease them out of poverty.

The activists and nascent "nationalists" did not stop with tract writing, however; they took their case, with the help of a number of British journalists such as William Digby, directly to the British people. Using the extreme conditions of famine in India as justification, many prominent Indians came to England and went on extensive speaking tours both to educate the British populace about famines in India and to raise money for the Indian Famine Relief Fund, which was founded—with the kind of irony that only empires can muster, at Lytton's sumptuous durbar in Delhi in 1877—in the midst of the Madras famine of 1876 to 1878.[14] These speaking tours and journalistic accounts continued into the century's end, and yet for all of the best intentions, the message became ironically and hopelessly misperceived. As many British citizens—ranging from lords of the nobility to textile factory workers—reached into their pockets to donate money for famine victims in India, and as new notions of responsibility entered into common parlance both in India and in Britain, many if not most Britishers began to associate India with famine and mass poverty. Even William Digby's scathing account of British famine policy during the Madras famine of 1876 to 1878 based its critique on the fact that the British had not done all that was expected of them to save the people of India from famine.[15] The critical voices against British rule, and the invectives that blamed the colonial administration for causing the famines of India, were slowly transformed into cries of indignation that the British had not done more to save India from famine and extirpate it from the Indian landscape. The message now seemed to be that India needed British rule more than ever, since no "Indians" seemed willing to take responsibility—why else would the Indians be asking the British for help? Instead of arguing that British rule had impoverished India, now the argument had become that British rule was necessary to ameliorate India's poverty. Responsibility for the famines lay in Indian hands, or at least in its landscape, but responsibility for saving Indians from these famines lay in British hands. Out of the nascent "nationalist" attempt to criticize British rule in India through the issue of famines and famine relief was born the powerful and persistent idea of foreign aid.

The issue of responsibility was also prominent in the so-called vernacular press of India. The Tamil newspaper *Dinavartamani* (8 August 1874) blamed the famines on the selfish behavior of Indian merchants and asked the British government to pass legislation against their price scheming. Another paper, *Vetticodeyan* (21 October 1876), published an editorial that echoed the indictment of Indian merchants and also gave a new twist to the call for the British to provide famine relief by stating that "famine interferes with the joy with which we looked forward to your being proclaimed Empress of this land." Perhaps the most intriguing statement came from the Telugu daily *Shams-ul-Akhbar,* which praised European efforts of private relief and charity for India and the symbolic charity of the Delhi durbar in

1877 and then decried the fact that their "co-religionists" had gone off to Delhi in the midst of the famine. A retrospective article that appeared in a Tamil monthly in 1912 summarized the many arguments in circulation about the causes of famines in India—including the arguments of Dutt and Naoroji—but then complained that the rhetoric of these debates had become more important than the famines themselves. Likening the rhetorical machinations to the myopic and self-centered jockeying for government positions with complete disregard for the functioning of the government or the welfare of the citizens as a whole, the article went on to say of the debates on famine causality that "there is some truth in every one of them, but the great task still remaining is to study [the situation] and resolve on a plan of action for famine prevention. No one is there to do this—they are simply waiting around to talk." [16]

What all of these narratives represent is a reorientation of the moral, social, and economic landscapes of south Asia. We can see this reorientation in progress by looking at the unsettled and uncertain interpretations of famine causality and moral responsibility. There were many for whom British responsibility in causing famines in India was not self-evident, which suggests the possibility that local understandings of famine causality were very different from nationalist ones. Multiple narratives make it exceptionally difficult to pinpoint common perceptions or public sentiment, and one of the alluring qualities of a master narrative is that it can smooth over the inconsistencies in local versions and present a unified, "orthodox" representation. Of course, in doing so, master narratives are increasingly removed from the actual event they purport to represent, something that is at the base of the contemporary tendency to mistrust them. Yet simply because we mistrust the master narrative does not mean that local versions are necessarily truer because they are closer to the source. What we have are simply more, albeit smaller, narratives in restricted contexts. And what is worse, without the generalizing, overarching force of the master narrative, we lose any sense of structure or any indication of a recognizable pattern. The question to which I shall return in the conclusion is whether one can find a sense of "justice" without the normative coherence of a master narrative.

As the famines in India slowly began to fade after the century's end—causing many to rethink the claim that British rule was necessarily responsible for previous famines—attention shifted to what was perceived as the chronic conditions of hunger and malnutrition among the "poor, starving" Indian population.

Poverty

After the turn of the century, famine was gradually removed from the physical landscape and embedded in the population that inhabited that landscape. It was as if some sudden wind blew up the desiccated soil of the landscape and breathed life into it, making up a new type of person: the poor, hungry Indian (or south Asian). The "poverty-stricken" and "hungry masses" in India were seen, in essence, as a mass of individual famines, spreading out to the borders of the national landscape; India was transformed from a land of periodic famines into a land of chronic hunger within a relatively short period of time.

It is difficult to say what caused this change in perception, or to say with any certainty whether Britain caused India's poverty or merely "discovered" it in the same way that Charles Booth and Henry Mayhew "discovered" the poor of London in the nineteenth century. One could argue that India's poverty, and responsibility for it, came to be an issue only as the poor of England became comparatively wealthier; there is some truth to this, but it skirts dangerously close to the idea that Indian poverty was merely a case of "relative deprivation" when in fact the deprivation was often severe and absolute. Alternatively, and more cynically, one could argue that famine relief had provided convenient justification for the expansion of British rule in India, and with the famines gone, something else had to be found to provide equally valid justification. That something, of course, was the amelioration of India's newfound poverty. Yet the question of amelioration was centered not necessarily around poverty per se but rather around the ideas of charity and philanthropy. Here the implicit question was this: Why were Indians always looking to the state to help them out of their suffering, rather than to each other? During the famine of 1876 to 1878, for instance, the government of Madras revised its guidelines for dispensing charitable relief when it was felt that Indians rich and poor were coming to procure free grain from the state and ignoring their civic and social obligations to one another.[17] In a book published in 1901, E. Washburn Hopkins, professor of Sanskrit at Yale University, searched through all of the so-called ancient texts he could find and reached two conclusions: the first was that the idea that famine in India was somehow related to British rule was "simply preposterous," and the second was that the "history of the native's dealings with his poorer brother" revealed a historical lack of charitable sentiment in Indian society.[18] By this argument, the difference between Britain and India was not necessarily a matter of wealth, but rather that the British took care of the poor, whereas Indians either did not or could not.[19] Once again, we are back to the idea that outside help is needed for a "poor and helpless India."[20]

Whereas previously the issue of responsibility had centered around the question of who had caused the numerous famines in India, now the question was expanded: who had made India poor? Poverty, like famine, was something that had existed since as far back as anyone cared to reach into Indian history, and yet the issue of causing poverty, of attributing responsibility for the impoverishment of a person or, in this case, of a nation, was an altogether new phenomenon. Opponents and critics of British rule once again mobilized fictional pasts, arguing by supposition that if Britain had made India poor now, then it must have been rich in the golden days before colonialism. They argued then, as many still argue now, that Britain's wealth was pilfered from the coffers of India's golden past.[21] Poverty relief was by this reckoning not the benevolent charity from the rich to the poor but a taking back of what was rightfully India's in the first place.[22]

Social and charitable service, then, and the simultaneous development of agriculture were phrased by critics of imperial rule in terms of reviving the golden past that the British had either stolen or hidden. The adoption of scientific agriculture was not a British invention, it was argued, but had been advocated in the *Tirukkural* and other "traditional" texts. Yet all of this dovetailed nicely with British efforts—and the efforts of many "enlightened" maharajas and Indian economists—to intensify efforts to develop India's agriculture and reform its social structure. The introduction of cooperatives and village *panchayats,* for instance, was nothing other than the introduction of new ideas in the guise of tradition. And like their opponents, the British were also mining the historical archive for evidence that their schemes were not "foreign" but only a revival of the past. Charles Benson had collected a number of Tamil proverbs related to agriculture in 1908 in the hopes of showing that those who resisted agricultural development in Tamil country were resisting their own ancient culture. Benson "quoted" the *Tirukkural* to argue that if the land did not produce good yield, it was because the ryot stood aloof from his work:

The earth, that kindly dame, will laugh to see,
Men seated idle pleading poverty.[23]

Although Benson may implicitly have argued that India's (rural) poverty stemmed largely from the collective forgetting of the traditions that had formerly made India prosperous, others read the same proverbs and inverted Benson's argument by emphasizing India's former prosperity under more "indigenous" (non-British) periods of rule.[24] In any case, from the end of the nineteenth century onward, the market in proverbs expanded manyfold and became intricately linked to the dual challenges of rewriting India's past and developing its future.[25] In the course of addressing these challenges, the ideas of economic development, and the ideas of poverty and hunger, became "naturalized" into the political and ideological landscape of India.

There were others, both British and Indian, who argued that India had always been poor, and that if it was becoming poorer, it was because there were already too many Indians, and the number of new Indians was growing out of proportion to what the land could bear. Radhakamal Mukherjee, the "father of development economics" in India, had argued in his book *Food Planning for Four Hundred Millions* that in the ancient past, Hindus practiced birth control and had small nuclear families; overpopulation, according to Mukherjee, was the result of derelict practices introduced by the Muslims, such as polygamy and child marriage, and the fact that poverty is a cultural ideal for Hindus.[26] Mukherjee also decried another population "problem" in India that he felt was equally as important as overpopulation—namely, the problem of "mis-population," whereby higher castes, because of "dysgenic customs like rigid hypergamy and endogamy" reveal a decline in fertility while the "less literate and backward classes" become "more fecund" and "threaten to swamp the cultured stocks."[27] In 1944 Gyan Chand, professor of economics at Patna College, wrote that "India is a country of death and poverty-ridden, derelict people" and summarized a common sentiment about India's population as follows:

> Whatever the political future of India and the extent of power it gives her [*sic*] over her own affairs, she will not be able to solve the problems of poverty, disease and death as long as her people let the natural impulses have free play and children are born without any rational forethought.[28]

The problem of overpopulation, like the problem of poverty, had been firmly set—by parties on all sides of the debate—into India's economic landscape, where it has remained ever since.

In a general sense, what writers such as Chand and Mukherjee represent is a new effort to moralize economics, or to economize morality, by welding together moral passion and social commitment within the putatively disinterested and objective context of economics, and to use the technical expertise offered by the tools of that discipline to achieve social justice and political reform.[29] Once again, through a strange juxtaposition of arguments, the supporters and detractors of imperial rule ended up, in spite of their rhetoric, arguing for the same thing—in this case, for more agricultural and economic development. The argument was also filtering down into more mainstream journals and slowly spreading into common parlance and "popular culture." An article that appeared in the January 1920 issue of the *Journal of the Madras Agriculturist Student's Union* recounted the history of Indian agriculture from its origins to 1920 and noted that charity and agricultural progress had gone hand in hand in the "ancient past," but over time (the details are unclear), ancient tradition and custom stultified the minds of Indians, causing the decay of community spirit and national unity that had allowed colonialism to take root. The conclusion of the article is telling: "The remedy for poverty lay chiefly in the people raising [*sic*] by strenuous efforts to the level of other forward nations in point of education, scientific knowledge, co-operation and industry."[30]

CONCLUSION

In the end, the debate may shift once again away from who caused India's famines or who caused its poverty and toward the question of who is responsible for making the image of India as a poor, starving nation persist. The "poem" quoted earlier in this chapter, *The Famished Village,* ends with the supposed plea of India's ryots:

We cry—relieve us. Let us live. Do grant
This boon. For aid from thee we anxious pant.

Offensive and patronizing as it is, it is not that far removed in rhetoric and tone from the myriad ads that currently grace the screens of our televisions and the pages of our magazines and ask for contributions to help a starving child somewhere in India or in some other equally hungry and helpless "Third World" country. From a different angle, a recent editorial in the *Wall Street Journal* (22 August 1995) called on India to stop whining about its poverty. The editorial stated that India was in fact a relatively wealthy nation but used the image of poverty to avoid responsibility for taking care of its own poor, or for taking even rudimentary steps toward financial or economic reform. To give a quote: "We are still waiting for an Indian leader who will stand up and challenge the political class to admit its own responsibility for keeping India poor."

And yet it is not just the West that trades on these images. A 1970 report on the famine in Bihar stated that the "scourge of poverty and famine" will be eradicated from India only when people learn to limit their families.[31] If we can believe the rhetoric of India's development economists and government officials, or the rhetoric of the family planning slogans that are written on the backs of lorries and engraved on the national currency, the dual threats of poverty and famine are always lurking just around the corner. One member of Parliament from Andhra Pradesh who has written extensively on rural areas has proposed (as so many have proposed before) a poverty eradication program in rural areas through a vigorous village development campaign—a linkage that, as we have seen, first emerged in British colonial famine policy.[32] Elsewhere, the Andhra Pradesh Civil Rights Association has transformed the failure of the central or state government to provide famine relief (for humans and cattle) and economic development (in the form of roads and schools), and to prevent starvation in the 1992 to 1993 famine in Mahbubnagar, into a violation of the civil rights of Indian citizens, specifically in relation to the right to a dignified life and the right to well-being that is enshrined in the Indian Constitution.[33] The questions of responsibility and blame are political questions, whose answers may prove permanently elusive, and if we are drawn continuously to try to answer them, it is only because it is difficult to imagine what the modern history of India would look like if responsibility and blame were removed from the historical landscape.

It is also of some importance to examine why it is that questions of responsibility and blame are often viewed from the perspective of the "proper" relationship between a state and its citizens. During the years just after independence, when Nehru was at the helm of development planning, there was a belief that the proper role of the state was a pervasive one, and that citizens should look to the state (as owner of resources and controller of the environment) for assistance in times of need and guidance in times of prosperity. The moral underpinnings of current debates over economic liberalization in India, however, reveal that this view has for the most part fallen from favor, and the new position of the state is one in which citizens are increasingly supposed to look to one another (and not to the state) for support and guidance. Here, the state does not control resources but merely monitors them. The belief that the state was somehow a kind parent to whom the (childlike) citizens could turn in times of need has given way to the belief that it is time for citizens to move out of the house of the state and start a life of their own. Economic liberalization is therefore predicated morally on the cultivation of liberalized selves among the citizenry of India.

The final point to be made here is not that there is no truth, or that the rhetoric that surrounds the event of famine is only a "construction," and that therefore there is nothing we can do in the way of seeking justice. Rather, it is to suggest the possibility of seeking justice—and it is important to keep in mind that in relation to famine or the environment, justice is not an absolute value but the product of changing attitudes toward the natural landscape—even in situations where the truth, or "what actually happened," remains inaccessible or opaque. Any attempt to reconstruct what is inaccessible will ultimately tend toward a master narrative that will perhaps buttress claims for the rendering of justice but will do so at the cost of distorting the historical landscape beyond recognition. In other words, master narratives may help produce a sense of justice, but they do so at the same time they render that justice unstable. Much of this historical distortion has taken the form of scapegoating, of singling out certain groups or institutions (capitalism, colonialism, grain merchants, etc.), essentializing their characteristics, and then attributing all occurrences of misfortune to them in order to excoriate them from a putatively omniscient perspective. But scapegoating is not the same thing as justice, and the sense of closure that historical scapegoating has produced will only be undone as more evidence becomes available and we see that responsibility and blame are distributed generously everywhere according to indiscernible or at best vague patterns.

The question to which environmental studies in general, and the study of famine in particular, must address itself, is that if all the master narratives we have at hand are rendered "open" and incomplete, where in the landscape will justice be found? The narrative landscape, like the natural landscape, consists of multiple layers and regions that cannot be understood properly outside of the context within which they exist. Understanding how and why a famine is represented, or how and why narratives of blame and resistance are constructed, is part and parcel of the same morally charged process of understanding famine and poverty as purely economic events. The differences we see in the representation of, and attitudes toward, poverty and hunger in India in precolonial, colonial, and postcolonial contexts should *not* be explained in terms of the "triumph of freedom," the "rendering of justice," the "loss and recovery" of culture, or the rise of the nation-state. Rather, the differences should be explained in terms of different and ever-changing moral environments, only some of which relate directly to poverty, hunger, imperialism, and other "charged" categories and historical processes. The result, I would suggest, would be a narrative that is neither lamentation nor polemic, but something in between that allows for contradictions and injustices in unexpected combinations and in unpredictable places. Justice, rooted in uncertainty, will then become permanent in an ever-shifting landscape.

NOTES

1 By "master narratives" I mean the collective records that compose a larger chronological narrative of causality; that is, narratives that attempt to order isolated events into coherent, causal patterns based on chronological progression. Thus in general, narratives such as the "search for social justice," or, in the context of south Asian history, the "freedom struggle" against the British, constitute master narratives in the sense that they interpret and validate specific events based on the role they play in helping to achieve a predetermined, morally charged end point ("freedom," "justice," and so forth).

2 Samir Ghosh, *Pancasher manbastar silpe sahitye* (Calcutta, 1994) [Bengali], 11. "Living memory" is *tathya-smrtite* (literally, "in true-memory"). *Pancasher manbastar* is "the great calamity of '50," referring to 1350 in the Bengali reckoning (1943 by "Western" reckoning).

3 Baccan, *Bangal ka kala,* 8th ed. (Delhi, 1964 [1946]) [Hindi], 24.

4 Panchugopal Bhaduri, *Aftermath of Bengal Famine: Problem of Rehabilitation and Our Task* (Calcutta, 1945), 47–52.

5 Christopher Wrigley (1996, 6–11) has argued that in relation to Buganda, the oral histories that scholars have often tried to use to recover the "indigenous" past of Buganda are composed primarily of invented traditions, idealized practices, and mythical pasts.

6 Fukazawa (1982, 476) notes that during famines, "government as a rule had to remit the

revenue, and would import foodgrains from surplus zones to open government shops and free kitchens for the afflicted people." Nevertheless, in recounting the effects of the famines of the seventeenth century, he cites frightfully high morbidity figures—one million in Ahmadnagar in 1630 to 1631, two million in the Deccan in 1702 to 1704—which indicate the ineffectiveness or perhaps symbolic nature of such measures.

7 In *Purananuru* (a collection of ancient Tamil poems dealing roughly with "public" life), for instance, the poet Peruncittiranar (poem 163) praises his wife for redistributing surplus wealth gained from a magnanimous local chief (Kumanan) by feeding relatives and assuaging their "intense hunger" *(katum paci)*. The redistribution of wealth is a common theme and much-praised tenet of so-called precapitalist economies, and yet the more important question here is why the relatives were suffering intense hunger in the first place. This poem was not composed during a famine, so we must entertain the possibility that the "moral economy" described in Tamil didactic literature saw no contradiction in the persistence of "intense hunger." There is much evidence to support this. In the *Tiruvacagam* of Manikkavacagar, chapter 4, lines 28–29, hunger seems to have existed right from the creation of the world, when days consisted of "purification in the morning, hunger in the afternoon, and sleep at night" *(kalai malamotu, katumpakar paci, nici velai nittirai)*. The point here is that the "evidence" is ambivalent at best, and that there is no discernible pattern within the ancient archive that allows one to establish a link between then and now.

8 *The Famished Village, or An Appeal to Englishmen in India and England for Relieving the Famine-Stricken Ryots in India* (Bombay, 1876), 5.

9 The most often cited reason for the Bengal famine of 1770 was the confiscation of all reserves by the British to feed the military. Certainly there is some evidence for this, but there is also evidence of similar actions both in the "indigenous" wars in the Deccan in the seventeenth century and in the tactics of Tipu Sultan as he fought the British at the end of the eighteenth century. This is not to excuse such tactics but only to point out that they were not necessarily new and were definitely not uniquely colonial. Moreover, as McLane (1993, 205–6) has recently shown, at least some of the misery was caused by local practices, and some peasants even used the disorder of the famine as an opportunity to relocate their villages outside the bounds of former socioeconomic links ("moral economies") to escape oppression.

10 The quotation is from the *Review of the Madras Famine, 1876–78* (Madras), 13. A similar opinion can be found in James Caird and H. E. Sullivan, "Famine Administration," India Office Records (IOR) L/E/5/66 (8 May 1880).

11 *Irrigation Works of India* (London, 1905), 320–21.

12 *History of the Periyar Project* (Madras, 1899), 7, 147; italics in original.

13 *India in the Victorian Age* (London, 1904), 78.

14 The Indian Famine Relief Fund, a showcase of private initiative and public (civic) philanthropy, was also designed to overshadow direct government relief, as this was thought to contribute to "excessive dependency" on government welfare and to discourage the charitable impulse among Indians. In the first season of the Madras famine, the government of Madras bought stocks of grain and transported them to areas of scarcity (or areas of high prices); at the Famine Conference during the Delhi Durbar, the Madras government was

chided for this and told not to purchase or import grain in any way that might interfere with the normal functioning of the market. See *Review of the Madras Famine, 1876–78*, 27.

15 William Digby, *The Famine Campaign in Southern India* (London, 1878).

16 "Intiyavil untakum pancangal," *Pilaikkum vali* 4 (April 1912): 161 [Tamil].

17 Tamil Nadu Archives, Government Order 2847 (Revenue), 24 September 1877. "And further to qualify for this [charitable] relief these persons must be without well-to-do friends or relatives on whom their support would ordinarily and rightly devolve." The sudden appearance of "well-to-do" Indians seems incongruous with the general talk of poverty, but as this quotation shows, they are mentioned only to highlight a supposed dereliction in their civic and social obligations.

18 E. Washburn Hopkins, *India Old and New* (New York, 1901), 232, 248.

19 Similar opinions were expressed during the Great Irish Famine. Cf. Alexander Somerville, *Letters from Ireland during the Famine of 1847*, ed. K. D. M. Snell (Dublin, 1994), 95, letter dated Longford, 5 March 1847: "Public opinion and public generosity in England are far in advance of public opinion and public generosity in Ireland."

20 There were, of course, attempts to encourage self-help organizations and Hindu charities, especially after the *swadeshi* campaigns of 1905 to 1906. Many of these, however, are merely attempts to take the values and beliefs implied by, or enshrined in, the British Famine Codes and give them an "indigenous" pedigree. Narayana Srinivasa Rajapurohit, *Danadharmapaddhati* (Dharwar, 1910) [Kannada], for instance, outlines the appropriate manner of establishing Hindu charities according to the principles supposedly contained in ancient Sanskrit texts such as the Bhagavad Gita or the Manavadharmasastra. Yet even here there is a case made for "discriminate" charity to protect against free riders, substituting Hindu criteria for what would otherwise be British or Christian norms. T. S. Cuppiramaniya Ayyar, *Celvavirutti Vilakkam* (Madras, 1915) [Tamil], 172, cites indiscriminate charity *(vivekamarra tarumam)* as one of the causes for India's poverty.

21 Greenough (1982), for instance, has argued that popular and nationalist sentiment in Bengal in relation to the occurrence of famines, particularly the ghastly famine of 1944 to 1945, gravitates around the idea that "outsiders" and "foreigners" demolished Golden Bengal *(sonar bangla)* and left only poverty and destitution in its place.

22 At a speech at the Plumstead Radical Club (21 July 1900) in England, Dadabhai Naoroji stated that Britain must not only pay the full costs of all famine relief and treatment for epidemics but also restore India to its "normal industrial condition" and restore its former prosperity. In *Poverty and Un-British Rule in India* (reprint, New Delhi, 1988), 573.

23 Department of Land Records and Agriculture, Madras: Agriculture Branch, vol. 2, bulletin no. 34 (1908): 21.

24 Pingali Nagindra Ravu, *Janmabhumi* (Masulipatnam, 1922) [Telugu], a short pamphlet of nationalist verse, contrasts the former greatness and prosperity of India with its present poverty and degradation in order to enlist nationalist sympathy and gather support for the noncooperation movement.

25 A sampling of writings on, and collections of, proverbs and folklore in southern India during this period would include *Samatisangraha* (Belgaum, 1906) [Kannada]; Narasingham Rao, *Proverbs Handbook—Kanarese* (1906); *Noti bodanegalu* (n.d.) [Kannada]; *Tatapata Haniyappa*

(Mangalore, 1922) [Kannada]; V. T. Sankunni Menon, *Parancol kathakal* (Calicut, 1932) [Malayalam]; *Parancol malika* (1906) [Malayalam], with English renderings; Government of Madras, *Cuttame cukam tarum* (Madras, 1958) [Tamil], a collection of Tamil proverbs for use in schools; P. Mutaiyapillai, "Tirukkural mulamum uraiyum," *Velalan* 1, no. 8 (1922): 242 [Tamil]; "Kavarnmentarin tuntuppiracuramum, manutarma sastiramum" (The Government's Pamphlet and Manu's Institutes), *Miracudar* 1, no. 5 (February 1935): 135–43 [Tamil and English]; "Palamolikal," *Kirusikan* 2, no. 7 (1910): 108 [Tamil]; "Pracin Vyavasayamu," *Vyavasayamu* 1 (March 1909): 94 [Telugu].

26 Radhakamal Mukherjee, *Food Planning for Four Hundred Millions* (London, 1938), xiii; also *The Foundations of Indian Economics* (London, 1916), 457.

27 Mukherjee, address at the Malthus Centenary (Lucknow), quoted in *Landholder's Journal* 3, no. 8 (May 1935): 728.

28 *The Problem of Population* (London, 1944), 4, 6.

29 This combination of technical expertise and social commitment still continues in south Asia and can be seen, for instance, in the works of Amartya Sen, to name one among many.

30 G. Jogi Raju, in vol. 3, no. 1, 28–46.

31 *People Can Avert Famine* (Delhi, 1970), 11.

32 Yalamamcili Sivaji, "Pallelu purogamimcanide pedarigam tolagadu," *Annadata sukhibhava* (Guntur, 1994) [Telugu], 21–24.

33 Andhrapradesh paurahakkula sangham, *Brahmajemulla brndavanam: Palamuru jilla paurhakkula mukhacitram* (Mahbubnagar, 1993), 13–14 [Telugu].

Sumit Guha

Economic Rents and Natural Resources:
Commons and Conflicts in Premodern India

The editors of this volume begin their introduction with an attack on the uncritical application of rigid binary categories to the understanding of environmental problems in India and elsewhere. They point to the need to "unpack" discursive structures such as the binary opposition of traditional conserving cultures with rapacious modern ones. A growing body of scholarship has sought to analyze the diverse ways societies have, in fact, spared and conserved resources for long-run use, and the conditions under which this has been achieved (Ostrom 1990). But before we can begin such an analysis, it is necessary to clear away extant misunderstandings. This chapter may be seen as a historical contribution to that process. It is therefore specifically focused on significant writers on south Asian environmental history, notably Madhav Gadgil.

The specific assumptions critiqued are argued by Gadgil, but they are widely present in the works of many authorities. For example, the corollary of the simple opposition of traditional and modern is that any opportunistic behavior among "traditional" communities can only be explained by their recent infection of modernity. So the *Second Citizen's Report on the Environment* declared that it was only now that groups that had lived for ages "in total harmony" with the forests wanted to sell them off as fast as they could (CSE 1985, 376). Such understandings of the past lead to unrealistic programs for the future. All too often, sweeping endorsements of little-studied "traditional" institutions have been substituted for their actual analysis in the work of (for example) Vandana Shiva, who writes that in South Asia,

> for centuries, vital natural resources like land, water and forests had been controlled and used collectively by village communities, thus ensuring a sustainable use of these renewable resources. . . .
>
> Colonial domination systematically transformed the common vital resources into commodities for generating profits and the growth of revenues. (Shiva 1991, 14)

This belief in turn leads to declarations such as that made in the Ninth Plan paper prepared by the government of India: "There is a symbiotic relationship between the tribal communities and the forests in which they live. The local tribal communities will be fully involved in the management of the forests" (GOI 1997, 65–66).

It must then be evident that a clearer understanding of interactions between humans and nature in the historic past is not without relevance to the present. This chapter tries to develop such an understanding. It begins with a critique of the only coherent and informed model of the historic past in South Asia—that developed by Madhav Gadgil and his collaborators—and goes on to propound an alternative view. The new perspective takes explicit account of the agrarian and hierarchical nature of Indian society and then tries to explain the regime of the commons within that context. It points to the centrality of power relations in determining the uses of the landscape. It thus goes beyond Gadgil, who had recognized social differentiation but sought to "naturalize" it via the analogy of distinct biological species occupying noncompetitive niches (Gadgil and Guha 1992, 105).

Gadgil and Guha (1995) also perceive a fundamental social divide between "ecosystem people" and "omnivores." The former are said to depend on the natural environments of their own locality to meet most of their material needs, whereas the latter partake of a global market in commoditized resources. Variations in their patterns of resource use then arise from this dichotomy: localization enforces prudence, and mobility permits extravagance (Gadgil and Guha 1992, 94, 106).

Unlike Shiva, Gadgil and Guha base themselves on their own pioneering work in ecological history, which presented the first (and, so far, only) coherent model of how natural resources may have been conserved and managed in premodern south Asia: a model designed to explain the interaction of humans and nature over some two millennia. Like all such pioneering efforts, it is bound to attract criticisms, and later scholarship will revise various aspects of their work; but such smaller local enterprises would be impossible unless preceded by their bold global venture. This chapter will have occasion to differ from them but is nonetheless deeply indebted to their initiative.

Gadgil and Guha face up to the fact that dense human populations such as those that inhabited much of south Asia from early historic times inevitably make large demands on regional ecosystems, and the authors in fact postulate that scarcities were encountered by about the middle of the first millennium C.E.; this led to what they term "conservation from below." The key institution in this was the "caste-based village society," and more generally, a social system that "very often ensured that a single caste group had a monopoly over the use of any specific resource from a given locale." This in turn ensured prudent resource use, for "small numbers of people linked together by bonds of kinship, and by a common culture, have had a monopoly over specified resources in specified localities" (Gadgil and Guha 1992, 94–95).

Implicitly, therefore, this formulation also addresses the vital question of the checks to overexploitation and answers that caste communities policed their boundaries against outsiders to prevent them from denuding local resources. The risk of similar misconduct by members of the community was, on the other hand, eliminated by a shared cultural outlook and also by a common sociobiological interest. The system of caste endogamy, Gadgil believes, would ensure that insiders were closely related and would have an interest in handing on an unimpaired resource to their descendants. Resources widely used by almost everyone, such as firewood, were, Gadgil states, controlled and managed by castes of village servants such as the Mahars of Maharashtra. The first generalized version of this model was presented by Gadgil and Malhotra (1983).

This model was built on the basis of extensive fieldwork over nearly two decades with both peasant and itinerant communities of western India but is supported by relatively little historical evidence. Such efforts run the risk of identifying recent developments as vestiges of ancient institutions—in particular, the assertion of specific claims to biotic resources that might be a reaction to nineteenth- and twentieth-century shortages rather than a relic of the previous millennium. We have records of the formulation of such claims. For example, the formerly densely wooded Thana district adjoining Bombay city began to experience local shortages of wood under the impact of improved communications and government controls from the 1860s. Villagers soon became aware of their interests. Deposing before a government commission in 1885, Hira bin Dharma, the tribal (Malhar Koli) headman of the village of Dabhon in Dahanu subdivision, stated: "We object to outsiders using our jungle as that diminishes our supply. As to *karvi* [reed] the supply is practically unlimited, so we do not mind outsiders taking them." If the villagers went on to succeed in excluding outside claimants from the local forests, field investigators a century later might well find as "immemorial custom" that villagers had an exclusive right to wood, but that nonvillagers could only take reeds. The headman's objection, we should note, was to free access; outsiders paying cash were less objectionable: "removal of head-loads [of firewood] for sale reduces our fuel-supply but we must continue to sell as that supplies us with money wherewith to buy our condiments" (GOB 1887, 2:84–85). Scarcity as well as the fungibility of scarce resources was understood without the benefit of Marshall's *Principles*.

Nor was this an isolated instance. By the 1880s, shortages were already modifying customs. When an official visited Belkade, a village in Kolaba district, he was told that the villagers had formerly gotten their wood loppings from Dhavar, "but the people of that village objected now to their taking it." The customary right *(vahivat)* of nonresidents to forest produce was itself open to dispute, and the forest settlement officer described how in some parts of the district, "you find that while the residents concur that no such *vahivat* amongst non-residents exists, yet some non-residents claim such a *vahivat* for their village, and some do not" (GOB 1887, 2:308, 47). Thus custom and usage might not be as fixed as Gadgil assumes, and peasant communities were far from being prisoners of their past. If this is once admitted, then the presumption that modern fieldwork reveals millennial custom becomes unsustainable.

This criticism may certainly be made of the mutualism between different caste communities suggested in Gadgil and Malhotra's study of the cultivators (Kunbi) and pastoralists (Gavli) of the Sahyadri ranges in the 1970s. The authors claimed that in the past, the Gavlis cultivated a little but largely got their cereals from the Kunbis, whom the Gavlis supplied with dairy products. This pattern was (they believe) disrupted by dam construction and population growth in the 1920s, after which, faced with a shrinkage of available territory, the Gavlis "began to intensify the shifting cultivation of the hill plateaus and upper hill slopes" (Gadgil and Malhotra 1982). But this disrespect for traditional occupational boundaries seems to have been equally traditional; it is found as long ago as the 1850s, when an early conservator of forests wrote of the need to regulate swidden because of

> the competition by the Gowlees, or wandering buffalo-feeders, with the more settled population in cultivating by destruction of jungle [i.e., swidden], the slopes and ridges of the ghaut hills. . . .
> The complaints of the villagers regarding these Gowlees were loud and frequent, not only because they interfered with the spots which they themselves had set aside for hill-cultivation [swidden], but in respect of the reckless manner in which they destroyed young trees within the village limits, and that they competed with the fixed cultivators on terms of inequality. (Gibson 1861, 4)

It is even likely that the government did intervene in the matter and thus helped shape the mutualism of Kunbi and Gavli that Gadgil's informants recollected a century later. Such social engineering by the Forest Department was reported from Kanara district (now in Karnataka) in 1921: "Several familes of Gowlis were brought in from Mysore. A plentiful supply of milk and ghee should help local villagers" (GOB 1921–1992, 34). Local villagers in turn provided conscript labor to the Forest Department, and their villages served as bases for its staff.

When one thus filters out possible anachronisms, the only tangible historical evidence that one finds in Gadgil turns out to be a reference to Atre's monograph *Gaon-Gada:*

> In a fascinating record of pre-British Maharashtra, Atre (1915) mentions that the Mahars also had the function of preventing any unauthorised wood-cutting in village common land. Additionally, they

had to harvest and deliver all wood needed by village households. . . .
[So their] interests would obviously lie in maintaining harvests from
the village common lands at a sustainable level. (Gadgil and Guha
1992, 94–95)

Atre himself, however, merely lists among the Mahars' duties "at night,
patrolling the village and preserving the village forest and trees"; there is
no mention of their supplying the entire village with wood. In fact, the
same page states that the Mahars were obliged to furnish fodder and fuel to
"important people and officials" who camped in the village (Atre 1915, 50).
Lesser folk evidently had to fend for themselves or perhaps purchase cow
dung cakes from the poor women who made and sold these (Manwaring
[1898] 1991, no. 1071).

Atre was an experienced colonial official, and much of what he wrote
was based on personal experience during the decades that preceded World
War I; he cites no pre-British record for the statements quoted earlier, and
we may take them to reflect the official view prevalent in the later nineteenth century, rather than anything earlier. Indeed, it is quite possible that
the duty of guarding trees was a new one, imposed as a consequence of
the creation of a colonial forest administration in the second half of the
nineteenth century. This hypothesis is suggested by the fact that I have not
found it included in the lists of duties prepared by inquiring officials in the
1820s at the very outset of colonial rule in West Maharashtra. W. H. Sykes
toured western Maharashtra between 1825 and 1829 in the capacity of statistical reporter to the government of Bombay and took particular pains to
discover the Mahars' roles in villages "where old customs may be supposed
to remain unaffected by the change of government." He noted a widespread
obligation to supply wood and grass to government officials but says nothing of other villagers. Intriguingly, he stated categorically that "in no instance . . . did I find them [Mahars] performing watch and ward for the
village"; this was the duty of Bhil or Ramoshi watchmen. Atre, however,
lists it as a duty of the Mahars—it is likely, therefore, that he was describing
the late colonial rather than the Maratha system (Sykes 1835, 226–28; Atre
1915, 50). So it appears that the British government relieved the Mahars
of certain obligations, such as forced labor away from home, but also imposed new ones, such as guarding the village and its depleted woodlands.
The latter was not a traditional duty, performed through the centuries, and

it follows that the mechanism for ensuring the sustainable use of the wood-lands proposed by Gadgil—that the same families had the hereditary man-agement of them—would not have worked. What, then, was the regime of the commons, if any, in premodern Maharashtra?

DEFINING BOUNDARIES THROUGH CONFLICTS

Gadgil and his collaborators have thus generated a frame of reference and raised all the important questions, but they have not really succeeded in providing an adequate historical foundation for their views—nor indeed would it be fair to expect them to have done so much at the very outset. However, the study of ecosystems does require a long perspective, as does our understanding of ways in which they have been managed by humans; and this chapter attempts a small contribution to the study of this problem via the abundant documentation available for precolonial Maharashtra.

A major strength of Gadgil and Guha's work is that the fact of scarcity ("resource crunch") and conflict over resources is built into the structure of the argument. This offers a refreshing contrast to the ineffable harmonies and mystic communions postulated in the work of Shiva and her ilk. This chapter searches for the evidence of controls and conventions by looking at the conflicts that perturbed a small town and the villages of its hinter-land in eighteenth-century Maharashtra. It does so in the belief that the study of conflicts is our best point of entry to an understanding of the actual social enforcement of rules and norms. After all, if we are to speak of a conscious regulation, then there must be occasional tendencies to in-fringe—if the latter are lacking, then obviously there is no need to regulate at all, and there is no "regime" to govern that activity. If, for example, an eighteenth-century fishery were to order that no one use nets more than fifty kilometers in length, this would obviously not be a rule that needed enforcement—it was impossible to make and haul such nets at the time. If this rule were the only regulation imposed, then we would be entitled to say that fishing was unregulated, and the seas constituted a free com-mon. Moreover, in a society largely governed by unwritten custom and usage, it is offenses that leave records; general statements of rules are few (see S. Guha 1995, 105–7).

So this chapter searches a "traditional" society for evidence of the sort of natural resource partitioning on community lines that Madhav Gadgil feels characterized precolonial India, and reports on both the attitudes and the disputes of country folk in rural India two hundred years ago. It is based largely, but not exclusively, on a collection of several hundred documents published by Oturkar in 1950.

THE ECOLOGICAL AND POLITICAL BACKGROUND
TO THE REGIME OF THE COMMONS

Before we begin our study of these documents, we need to outline the social and political regime in which they were generated. The small town of Saswad was a part of the territory of the Maratha kingdom notionally ruled by the Chhatrapati resident at Satara, but the town was in fact governed from Pune, a few miles north of Saswad, by the hereditary ministers known as the Peshwas. Most of the open country was occupied by nucleated villages, each surrounded by cultivated fields and areas of scanty open woodland and pasture. Denser forests were found in the Sahyadri ranges to the west. Some of the peasants (termed *mirasdar*) in each village would have hereditary rights to their lands, and most village functionaries were hereditary holders of their offices. Hereditary office was known as *watan*, a term best translated as patrimony (see Fukazawa 1991). Caste, custom, and the watan were, as Oturkar has remarked, the three central institutions of the society of the time (Oturkar 1950, 7). The documents in Oturkar's collection originated in the office of the hereditary registrar and accountant *(deshpande)* of Saswad subdivision and reflect the tensions and disputes in local society. Saswad subdivision lay a few miles south of the important city of Pune, occupying the plains of small valley, closed off by the Sahyadri range to the west, and opening into the flat open lands of the Dakhan plateau to the east. The strategic hill fort of Purandhar was located in the hills overlooking the plain, and Saswad was a subdivision of the lands of Purandhar *taluka*. The stable though semiarid climate permitted the unirrigated cultivation of millets, oilseeds, and wheat. Vegetables, cereals, and a little sugarcane might be grown under irrigation, either from hill streams or from wells. In 1825 to 1826, when a comprehensive census was taken at the instance of W. H. Sykes, Saswad contained 7,088 inhabitants, living in 1,541 households; 747 of these were classified as "cultivators"—meaning landholders directly assessed to tax. There were also 106 households of shopkeepers and traders, and 59 weavers and various other craftsmen.

Though this was primarily an agricultural economy, trade was not an insignificant activity. There were 1,748 plow bullocks and also 312 pack oxen (MSS Eur D. 148, fols. 148ff.). Apart from the fixed shops, periodic markets and fairs also offered opportunities for commerce and linked the villagers with regional market networks. This would also have been true of the later eighteenth century, to which our documents refer.

If it was not economically isolated, the village was not politically isolated, either. Villages quarreled over boundaries, lineages feuded over hereditary posts, village servants neglected their duties, and individuals insulted or injured their fellows—all these issues were frequently carried up to the district administration, or even to the Peshwa's court at Pune. Thus, for example, in 1722 or 1723, Rayaji Katka, resident of Murti, beat his nephew; the latter complained to the district officer at Supe, and Rayaji was fined seven rupees (PT, vol. 14, fol. 77). Central authority was present, therefore, not only in intravillage but also in intrafamilial affairs.

As a consequence, even seemingly minor issues appear in the records. For example, a local Brahmin and the headman of the village of Chabli quarreled in 1819 over which of them was to precede the other in putting on perfume after a religious ceremony, and a full inquiry into usage in neighboring villages was ordered. Or again, we have papers recording charges and countercharges made over the issue of whether vegetable vendors were liable to compulsory grass cutting for the hereditary officials of Saswad (Oturkar 1950, 58, 45). Any hereditary right was of great significance; for example, an unnamed Brahmin possessed the right to inform the local Mahars of the auspicious moment for marriages and presumably collect some fee for that service. Another Brahmin, Vitthala, enemy of the former, gave out this information and was rebuked—"you destroy the *watandar*'s right, what is this?" Vitthala responded with abuse and was mulcted the considerable sum of twenty-five rupees (Oturkar 1950, 67). Similarly, a village headman claimed the shoulder blades of all goats sacrificed during the Dasahra festival; the ensuing dispute led to inquiries in the thirty-eight other villages of that subdivision (Khedebare) to establish the validity of the claim (Oturkar 1950, 51–53). Itinerants could also possess watans, so in 1722 the government confirmed the exclusive right of Sheikh Rustum and Sheikh Imam to exhibit leopards and bears in six subdivisions, and to receive one thirty-second of a rupee from each house (Oturkar 1950, 24).

Examples could be multiplied, but the foregoing should be sufficient to show that even minor village disputes could come on record, and that state authority was exerted within the family, the community, and the village. On the other hand, since many state offices were hereditary and liable to partition and sale, the state itself may be viewed as an aggregate of great families, and thus shading imperceptibly into the society that it dominated. Furthermore, bribes, gifts, patronage, flight, and defiance all served to mold the initiatives of central authority to individual or local needs—another issue that I have discussed elsewhere (S. Guha 1997).

To return to our theme, it follows that most issues of any significance to local society would leave some documentary record in the offices of lower-level functionaries such as the hereditary registrars of Saswad. Yet a review of the 218 documents in Oturkar reveals how rarely these "ecosystem people" quarreled over the spontaneously available resources of the land they inhabited. Furthermore, if caste divisions in south Asia were based, as Gadgil suggests, on the partitioning of resources between such local communities, we find no trace of such issues in the various inter- and intra-caste disputes that crop up in the records. We find two horticulturists *(mali)* being warned to withdraw the social boycott of Bhavan Mali that they had started by alleging that his mother belonged to another caste (Oturkar 1950, 125). So even membership of, or exclusion from, what is seen as the primordial community—caste—was subject to state interference. Such issues of money, honor, and standing—rather than the partitioning of natural resources—agitated rural society in our period. This was true even of the Mahars, the community whom Gadgil casts in the role of guardians and managers of the commons. A prolonged dispute with the village headmen was settled in the 1730s by specifying the gifts, fees, and perquisites that the Mahars were to have on various occasions, and stipulating that they were to receive (with certain exceptions) the hides of all cattle that died in the village (Oturkar 1950, 29–32). Two decades later, another dispute broke out, this time between the Mahars and Mangs who contested which community should carry an earthen pot during the exorcism of Saswad township. The enmity arising thereof persisted into the 1770s, when the Mahars struck work and their tasks were undertaken by the Mangs. The latter, however, began to find the burden of unpaid labor unbearable and finally withdrew at a strategic moment when an ox had died in the town: no caste Hindu could touch the carcass, and the whole vicinity was polluted by its presence. Frantic efforts were then made to induce the Mahars to return to work. Forty years later, in 1810, Mahars and Mangs were still disputing their rights (Oturkar 1950, 38, 44–45, 56). Similar issues of perquisites at festivals also came up in Pargaon in 1778, where the Mahars disputed the Chambhars' right to receive five shares of consecrated food at the Holi festival (Oturkar 1950, 49–50). Only once in Oturkar do we find a dispute over fodder: this was when grass cutters attached to the Peshwa's army invaded the houses of some temple attendants and carried off the grass they had stocked for their cows (Oturkar 1950, 21). In eighteenth-century Maharashtra, eminently human products—money, food, status, honor—and not raw natural resources were the focus of farmer, artisan, and laborer alike.

Why would this be the case? I suggest that the maintenance of control over woodland and fallow was, broadly speaking, of little interest to ordinary folk—their cattle could live on monsoon grasses and crop residues, and their mud huts, thorn fences, and dung fires could be sustained from a landscape of modified scrubland and savanna. Even if they had desired tighter control over this land, it is likely that they lacked the power and resources to police it—guarding their crops and homesteads was hard enough for them—and the scanty yields of "dead lands" (*pad zamin*) were left accessible to all.

LORDLY POWER AND NATURAL RESOURCES: INSTITUTING A REGIME OF THE COMMONS

Great men, chiefs and commanders, would not be content with such scanty yields, and so they both drew on biotic resources from further afield and also sometimes needed to manage and regulate local yields. So, for example, the large beams and rafters needed for palaces and temples had to be brought from considerable distances, typically from the hill forests or the western coast, for as W. H. Sykes noted in the 1820s:

> Everywhere within the Poona and Ahmed Nuggur Collectorates, nine-tenths of the timber used in architecture is teak wood and this is brought up from below the ghauts at enormous expence. (MSS Eur D.148, fols. 22–23)

This was an old pattern: when Tukoji Holkar wished to build an almshouse in Pandharpur in 1782, he bought the timber in Pune and planned to float it down the river from there (SSRPD, part 8, vol. 3, p. 215).

Apart from such occasional demands for large timber, there was the recurring need for firewood, and also for fodder to feed the numerous elephants, bullocks, and horses that transported great households, not to mention the cavalry chargers that maintained that greatness. These might sometimes be supported by sweeping off the peasants' stores, as in the case of the temple attendants already mentioned. In 1802 the Mahars attending on an English officer, Dipton (?) demanded two thousand bundles of fodder from a village near Burhanpur. Fodder was difficult to find and took time to procure, so the local official Parasrambhat was tied up and flogged. Other commanders were more provident if not less peremptory. Ibrahim Khan had charge of a unit of the Peshwa's cavalry; anticipating shortages, he issued an order to all village headmen in the tract east of Pune, demanding that all meadowlands should be reserved for his needs (Rajvade 1909, 449, 443). Nor were such demands very novel. In 1758 a Maratha officer posted at Ranthambor in Rajasthan was advised that another unit was joining him, and he should have an additional forty to fifty thousand bundles of grass cut and stored for them; in addition he should have an ample supply of firewood. These supplies would presumably have been extracted from village pastures and woodlands, probably by forced labor (Apte 1920, 55). Cavalry units would obviously claim priority over local requirements: in another news report of that period, we learn that men from a nearby military camp simply came and cut down all the standing crops, both rain fed and irrigated, in the village of Karathi in Khandesh. The villagers then thought of abandoning the village and settling elsewhere (Rajvade 1909, 453–54).

In less disturbed times, less destructive, though equally arbitrary, arrangements might emerge. A glimpse of how they took shape is afforded by a letter from the administrator of the township of Kadus to the Peshwa, written in March 1736.

> The honourable Rajshri Senapati's camp-followers go daily from the main camp to Talegaon. They turn elephants and camels into the fields, and they get into the irrigated lands and rob. The Lord (Peshwa) may command on this matter. Rajshri Mahadji Govind has been granted the village of Turakdi. He has just reserved its grazing lands; he beat (our) cowherds; to the north Rajshri Mahadji Govind has reserved the grazing, and that of Sayegaon is reserved by Raj-

shri Tryambakrao Mama. Where will the people of Kadus take their cattle to graze, from which forest will they fetch their wood? It is not possible to carry on the life of the settlement without touching the border tracts of the adjoining villages. The Lord is able to command. (SPD, vol. 30, p. 129)

This was not an unusual occurrence; thus in 1778 the headman of Kaloli, a village near Saswad, complained that a powerful noble, Jiuba Chitnis, had similarly closed some land in the adjoining village of Naloli. The rains had failed the previous year, and the only available water for the local cattle was in a ravine in the reserved lands. Chitnis's officer had beaten and threatened the local villagers when they took their animals there. The situation was aggravated by the arrival of the Peshwa's officer, Avji Kavde, with a large train, whose cattle also went to drink there. If this continued, the head-man reported that the peasants would be severely distressed (and might emigrate?) (Oturkar 1950, 7).

These documents bring out how inequalities of power affected the control of biotic resources: the small needs of local villagers and townspeople could be met under a regime of free commons, but closure was necessary when the gentry appeared on the scene. Such closures might be temporary, or they might become permanent; in that case, these lands would become the private or government "meadows" *(kuran)* that we often find in the records. Many of these formed the core of forest reserves in the nineteenth century. The creation of one such reserve was ordered by the Peshwa Bajirao in 1758. The order noted that the court often marched through the district of Karde-Ranjangaon and needed wood and fodder. So the local officer should find a village (preferably a partly cultivated one) assessed at four or five hundred rupees in tax, knock down most of it, allow a little cultivation to remain, and convert the remaining lands into a kuran (SSRPD, part 3, vol. 1, p. 282). Both the Peshwa's government and leading gentry families possessed numbers of meadows of this type, all probably created by excluding local villagers in the way described earlier (Gaikwad 1971, 84, 96).

Sometimes these reserves were made available to the peasants as well. For example, after the nizam of Hyderabad's army had burned many villages in the Junnar area, the peasants were allowed to take 100,000 bundles of grass as well as wood and bamboo from the Randhervadi kuran to re-build their homes (SSRPD, part 3, vol. 1, p. 251). These reserves had presumably earlier been selected for closure because of their productivity, but that they could contain resources unavailable in the village common lands also probably indicates that restricted access did increase production, by contrast to the more usual regime of free common.

CONCLUSIONS

We began this chapter by pointing out the pitfalls of generalizing from inadequate evidence, and it would be inappropriate to end by doing just that. It is necessary, therefore, to state that the findings presented here are based on a few hundred documents pertaining mainly to a part of western India—the region around the city of Pune, and the neighboring territory of Saswad. However, even this limited investigation shows us that any understanding of natural resource management that is based on the assumption that these were handled by static little communities inhabiting a stable universe far removed from the state and the market is fundamentally flawed.

The markets for goods and services and the markets in favors and exemptions were equally developed, and country folk needed to exploit opportunities and escape dangers in both of these if they were to survive. In this milieu, the village community was lucky if it could safeguard its grain stores and standing crops from the perils—*asmani va sultani* (from the heavens and from the kings)—that surrounded them; guarding the scanty produce of untilled land, savanna, and pasture was something that only great families and the state that was their collective patrimony could afford. This finding is not without contemporary relevance now that control of the commons by local communities is being advocated in South Asia and elsewhere, and its relevance lies in that it warns us that the simple retreat of *official* central power is not sufficient to create local autonomy or prudent management. Given the frequently evident mismanagement of resources by the modern state, it is obviously tempting to suggest that it needs to retreat and let the traditional communities take over. However, I hope that this study has demonstrated that the boundaries and functions of such communities were never uncontested, never "natural." Opportunistic transgression and arbitrary power were never absent from their working. People, things, and information move farther and faster now than they have ever done in the past, so the retreat of the state will by no means leave the "local communities" free to solve their problems. Indeed, it may simply leave them helpless before powerful individuals or nongovernmental organizations (such as mafias). A news report may be cited as an example—J. P. Yadav and Niladri Bose's "Under the Axe," on Saranda forest in South Bihar being raided by truck-borne timbermen (*Statesman* [New Delhi], 7 July 1996, supp., p. 2). It is important to realize that local autonomy must be accompanied by the vigorous exertion of state power, to defend both the geographic and legal boundaries of communities; only then can we expect the prudent management of resources by them.

Identities and Livelihoods: Gender, Ethnicity, and Nature in a South Bihar Village

One of the challenges for gender analysis of environmental relations is the need to steer between entirely materialist approaches to land and labor, on the one hand, and the absorption with identity, in which actual livelihoods and gender relations are neglected in favor of cultural representations, on the other. We need an understanding of both material transactions and the force of identity discourses in patterning actually existing social relations to see gendered environmental change in other than the polarized images of ecofeminism and populist environmentalism. Toward that end, this chapter looks at how *dikku* (Hindu incomers), *adivasi* (tribal indigenes), and *dalits* (low castes) [1] in a south Bihar village see themselves and others, how they envisage and justify their resource claims, what relationship there is between actual and idealized livelihoods, and how gender identities and relations are part of the legitimation of patterns of resource use and access, and of representations of ethnicity.

The separation of the agrarian, in the sense of rural production and livelihoods, from the environment, as nonarable nature, is problematic in a number of ways, as Agrawal and Sivaramakrishnan argue in the introduction to this volume. One of these is that such a separation feeds into the powerful tendency in radical Western environmentalism to insist that nature must be conserved for its own sake. Thus the term "natural *resources*" has become unacceptable to environmentalists because of the objection that nature should not be seen only as serving the interests of humanity.

Ramachandra Guha has argued, against antihumanist conservation imperialism, that "'Green missionaries' such as conservation biologists and their supporters are possibly more dangerous, and certainly more hypocritical, than their economic or religious counterparts" (1997, 19). Similarly, I would suggest that ecofeminists who insist that "ecofeminism must be concerned with the preservation and expansion of wilderness on the grounds that wilderness is an Other to the Self of Western culture and the master identity and that ecofeminism is concerned with the liberation of all subordinated Others" (Gaard 1997, 5) are potentially antiwomen and themselves closely aligned with the master identity of Western hegemony.[2]

Bioethical stances, and their implications for social justice in many poor rural communities, remain unchallenged by approaches that separate rural human livelihoods, primarily agriculture, from issues such as deforestation, since the identity of "farmer" is thereby separated from some of the other identities beloved of Western environmentalists such as "tree loving tribal/woman." However, these identities may exist within the same embodied individual. The plurality of identities lived by rural peoples and the contradictions these sometimes contain need to be recognized in the ways we understand environmental relations to avoid the romanticizing constructions shared by Indian populism (e.g., G. Sen 1992) and some Western environmentalism.

This chapter stems from research that aimed to focus on the natural resource perceptions of villagers, across a range of social divisions, in the context of shifting political realities and changing livelihoods.[3] It reveals how unsatisfactory are generalizations that present villages, communities, and women as unitary categories, how resource relations and representations are embedded in specific political and social contexts, and how instrumental relations with nature remain, although human instrumentalism, livelihood and political, can and does protect nature. If there is a need to dismantle homogenizing ideas of women and community, so too do we need to consider the levels and meanings of nature more carefully. Critical realists argue that nature is not only socially constructed but also a constraining physical universe. Kate Soper demonstrates the importance of distinguishing between nature at three levels: as a physical world, as what is managed and altered by humanity, and as an idea that demarcates social and cultural boundaries and is contested and defended in social movements (1995, 3). There is a danger that discourses of resistance that deal only with "nature as idea" will make faulty assumptions about "nature as managed human livelihoods," assumptions with flawed implications for development and environment policy and strategies.

From a different angle, it might be argued that understanding the politics of environmental relations and struggles requires greater attention to the detailed workings of existing livelihoods. Sherry Ortner (1995) has argued persuasively that understanding resistance requires more ethnographic thickness and a move beyond what she terms the ethnographic refusal that characterizes the field. She remarks that ethnographic thinness in studies of resistance results from a "failure of nerve surrounding questions of the internal politics of dominated groups and of the cultural authenticity of those groups" (1995, 190), and in this the internal politics of gender are particularly important. Ethnographic detail also reveals a complex and sometimes contradictory relationship between livelihoods and identity in which it is possible for cultural constructions of environmental identities to bear rather tenuous connections with actual livelihoods or resource relations. Constructions of ethnic identities can serve to assert what might be called "cultural entitlements" to nature, which do not necessarily either conserve nature or refigure gender relations in the manner that radical environmentalist and ecofeminist discourses (e.g., Eckersley 1992; Shiva 1989a) imply. In this chapter, we examine internal politics and authenticity in identities and gendered livelihoods, on a modest scale, for a single village of south Bihar.[4]

After locating the village of Phulchi, we will give a brief account of its political context, of changing constellations of power and their expressions in environmental relations. Then we look at the ways in which the main social groups of Phulchi perceive and represent their environments and each other. Finally we examine how real livelihoods connect with these ideas and how gender relations map onto this local landscape of ideas and actions, struggles, and silences. Phulchi village is located twenty-two kilometers south of the town of Giridih, close to the Usri River and in an area with extensive *sal* forests.[5] Rainfall is not particularly low, but its distribution is poor and erratic, and monsoons variable. Village agriculture is almost entirely rain fed, constrained by availability of arable land and relatively low crop productivity. Arable land is locally classified into five types along a topographical sequence and with varying soil properties, from *garha* (lowest land growing paddy), through *ajan* and *baad,* which are higher but also grow paddy, to *tanr* uplands where only extensive maize and crops like *kurthi* grow. *Barhi* plots are those around homesteads where intensive cultivation of vegetables, sometimes using well water, is possible.

The population of the village (1,207 in 1990) is made up of roughly similar proportions of the Bhumihar agricultural caste and tribal groups (mostly Santal with some Kohl), as well as a minority of Scheduled Castes, the Turi. The history of Phulchi is one of occupation and expansion by the Bhumihars, from north Bihar, at the expense of the tribal population.

In recent decades the nationalization of the coal mines, the closing down of hundreds of mines and the retrenchment of miners, and the acquisition of land for mining with little or no compensation, was met with protests both industrial and rural, such as the cutting down of teak plantations. Over the years, the Jharkhand movement has had both urban and agrarian elements, and since the early seventies Jharkhandist peasant protests have been most widespread and effective, involving forcible harvesting in which the Santal peasantry and their allies claimed the harvests from lands alienated by landlords and moneylenders. These protests were a regular seasonal occurrence in the Giridih area even before the Jharkhand Mukti Morcha was founded in 1972, and they were numerous through to 1976 and continue sporadically.

The Santal leader Sibu Soren came from Tundi (between Dhanbhad and Giridih), and people in Phulchi still speak of 1972 as the political watershed when Sibu Soren "awoke them" with his flute and began to change the balance of power within the village. The state government responded at some times with conciliation and ineffective development programs, and at other times with violent repression, and in 1980 split the area into new districts to try to fragment the movement. By the end of the 1980s, the Jharkhand movement had become reformist and ethnicist with cultural revival and separatism as their aims, whereas a more radical splinter group, the Lalkhandis (red land), rejects the regional alliance of all classes and conceives of a "community of labouring people" (Devalle 1992, 166).

Tribal cultures have been widely represented as less patriarchal than Hindu and as ecological in orientation, and indeed the Jharkhand movement defined itself in opposition to dikku oppressors in the same terms; tribal groups were said to protect their environments and to offer considerable freedoms to women. The Jharkhand Coordination Committee has defined the distinctive Jharkhand cultural identity as "harmony with nature; equality in society, including a relatively equal position for women; and collectivism in economic activities" (Kelkar and Nathan 1991, 22). Like other movements elsewhere in India, these have been powerful and effective images, claims, and identities. But where these discourses are held up in the light of ethnographic research, they have been seen as caricatures of adivasi lives, distortions of actors' own perceptions of their struggles, and as appropriations that make use of adivasis in the project of urban academics and activists seeking an authentic and indigenous critique of development (see Baviskar 1995 for an example).

Romanticization of tribal gender relations has been critiqued by Baviskar, who shows for the Bhilala how the politics of honor involve the control of women; and Kelkar and Nathan (1991) point out, for the Jharkhand, that gender relations enter into tribal peasant struggles through tribal men objecting to Hindu men abusing "their" women. The Kol rebellion of the 1830s in Singhbhum articulated retribution for the honor of tribal women as a major objective; in the Santal rebellion of the 1850s, witch-hunts against women as well as sexual abuse of young women by Santal leaders occurred; and the Munda uprising under Birsa appealed to *men* to reject antiwomen practices such as alcohol, dancing, and multiple marriages because they demeaned Munda identity to Hindus and Christians (Kelkar and Nathan 1991, 150–53). Women continue to be the means toward male ends, of maintaining adivasi honor and identity by defending women against abuse by non-adivasis, to assert adivasi equality with non-adivasi.

Clearly this proprietorial defense of women is an expression of tribal patriarchy, and Kelkar and Nathan are surely right to question its masquerade of meaning. Nevertheless it would be perverse to deny the real benefits for subaltern women, low caste and tribal, of the admittedly objectifying defense by tribal men of their male gender interests. In Phulchi, low-caste dalit families described the time before the Jharkhand movement in the 1970s as a time of violent exploitation in which women were especially vulnerable. Only twenty-five years ago, all newly wed girls had to spend the first night with their husband's landlord; they could not go to the river or forest without fear; and rape by Bhumihar masters of their female servants was commonplace. The Jharkhand movement has dramatically altered this scenario and created the conditions under which, for example, most dalit families created their own hamlet, away from the Bhumihars, with their

own wells, to keep their distance from the unwanted attentions of Bhumihar men. For their part, Bhumihars now attempt to represent themselves
as the victims of a government run for the benefit of adivasi and dalit, frequently citing as evidence their exclusion from government subsidies for
wells.[6] In particular, the chief minister of Bihar, Laloo Prasad Yadav, was
demonized as creating a situation where such groups had now "overtaken"
the Bhumihars in prosperity.

Caste relations in Bihar are complex; since independence, some upper
backward castes such as Yadavs have indeed achieved both economic and
political power through landownership, but such groups are hardly cham-
pions of the low castes. Indeed, some now participate in the caste-based
senas (armies) that since the 1970s have been involved in massacres and violent repression of Naxalite struggles (Bhatia 1997). Bhumihar discourses
of disadvantage lump together the power of previously lower caste groups
and what they represent as the pro-dalit and adivasi stance of the state at
the national level. The local reality, however, is that only some nominally
low-caste groups benefit from the Laloo government, which is actually
hostile to Jharkhandists and low-caste activists, and dilatory in carrying
through national-level programs for the disadvantaged. But although little
has changed materially, and the institutions of the state remain dominated
by high-caste partisans, the current political climate is one in which most
people in Phulchi see the legitimacy of Bhumihar domination challenged
as never before, through direct struggles, in the media, and through the increasing knowledge of rights and entitlements. How do these larger political processes connect with the ways in which environments are understood
locally by women and men of different social groups in Phulchi?

EXPLANATIONS OF ENVIRONMENTAL CHANGE

The first onion ring to be removed in trying to understand local explanations of environmental change is that of "outsider" perceptions of degradation.[7] The forests that surround Phulchi are denser than many in the region but nevertheless appear, to outsiders, considerably stressed. All the
trees tend to be of a similar age, and there are few saplings or mature trees,
species diversity is low, and all trees bear signs of heavy coppicing and pollarding. Soil erosion is evident in the bare upland tanr areas with dramatic
gullies of more than twenty-foot depth. The hamlet of Bithiya, where most

adivasis live, is located on these eroded uplands, whereas Bhumihar fields and dwellings cluster around the lowland rice fields on garha plots. Outsider perceptions of environmental degradation are not, however, shared within Phulchi, where the uniformity of the height of sal trees is seen by women as conducive to leaf plate collection. Gullies are said to have been there for a very long time, and where they are eventually blocked with boulders, the accumulating soil is enriched with manure and eventually cultivated. Sheet erosion is seen as part of processes of soil formation in which topsoil is washed down to the cultivable lands lower down. This is regarded with some ambivalence; on the one hand, soil *movement* is perceived rather than soil *loss*, but on the other hand, given the distribution of settlement and landownership, the Bhumihar controllers of lowlands are the winners from "erosion," but the adivasis on the uplands witness their lands losing fertility. What outsiders might see as anthropogenic degradation, insiders may see as natural processes and positive resource management. Setting aside the issue of whether environments are degraded in Phulchi, are they protected by adivasi sacralization of nature as G. Sen (1992) and many other ecopopulists assert?

It has been suggested that the sacralization of landscape and natural resources, by tribal peoples in particular, is part of a cultural predisposition to protect the environment (Banuri and Apffel-Marglin 1993). However, in Phulchi, all three social groups sustained beliefs and practiced rituals to appease forest spirits, to ensure the fertility of soils, to induce good monsoon rainfall, and to bless wells. These beliefs and practices did not necessarily affect people's actual resource management behavior or make them more environmentally concerned or aware. Some practices, such as the ceremonies to ensure a good monsoon, are features of high caste rather than tribal practice, and it cannot be said that tribal groups in Phulchi sacralize resources to a greater extent than castes, as the discourse on identity and environment suggests (e.g., Engel and Engel 1994). In some ways, religious observance around nature can even serve as a license to exploit resources, as Nagarajan (1996) shows in her analysis of the sacralization and desacralization of ground in the *kolam* practices of south Indian women, a license linked in the Phulchi case to the consolidation of hegemony.

In discussions about the changing quantity and quality of water resources in Phulchi, all men concurred that rainfall was poorer than in the past. The dominant local expression of this concern was the extremely large and expensive *yagya,* a religious festival to ensure good rains, organized by the Bhumihars. For this, Rs 85,000 was collected through subscriptions from every household to pay for priests giving continuous worship of Hanuman and Kali for nine days. The yagya has been held every year, but on a much smaller scale, since 1962 and is both a reflection of the perception of monsoon degradation and a powerful way of asserting Bhumihar social dominance.

More important than sacralization in Phulchi was a secular politicization of natural resources; thus there was considerable agreement among men and women across castes that forests had become depleted in recent years, but very different analyses of causes. Dalit and adivasi informants blamed construction activities and clearance for cultivation by Bhumihars as the main factor in deforestation; the loss of land and the loss of the forests are consistently related to Bhumihar exploitation, and no mention of other possible factors, such as population increase, figured in their accounts. The fate of the *mahua* trees preoccupies the adivasi community, since there are now said to be only forty to fifty of the trees left in the vicinity. These stand as a symbol of the past and their oppression; they are protected and valued but stimulate no efforts at regeneration, perhaps because they are no longer central to diets and livelihoods, which have diversified in other directions.

Bhumihar views on the state of the forest emphasize falling numbers of trees, the reduction in species diversity, the declining quality of the forest, and the loss of fruit trees in particular. They say that the trees have been removed "due to population increase," and they do not accept responsibility for forest clearances for arable land but claim that poor people are the cause of the problem because they are unable to buy timber or alternative materials for making houses. They also blamed the adivasis for exploiting the forest and said that from 1977 tribal people had begun to cut large quantities of trees with the justification that "we are Jharkhandi, everything in this area is ours." Thus a resource problem is perceived, but the analysis of the causes (and therefore implicitly the solutions) is filtered through antagonistic ethnic identities. Phulchi people invariably perceive resource problems through the political lens of their particular social position and perspective.

Perceptions of resource degradation in Phulchi are formed more strongly by ethnic identity than gender identity, and we found consistency in the views of men and women of each social group. Thus, for example,

Bhumihar women agreed with their husbands on the nature and causes of environmental change; they were as partial, as prejudiced, and as resentful. In addition they expressed, unlike their husbands, a certain scorn for "jungle" and rural life. Bhumihar wives with a little education were especially negative about being married in a rustic location, and some even welcomed the thinning of the forest as civilizing. Those who romanticize and essentialize "women's ways of knowing" (Banuri and Apffel-Marglin 1993), their sensitivity to nature, and their close involvement with resource management overlook this unpalatable reality. The consensus between Bhumihar women and men also reveals the limitations of approaches that emphasize women's identities and knowledges as determined by divisions of labor, for despite the striking differences in divisions of labor between, for example, Bhumihar men and women, they share many views on the causes of resource depletion. Opinions and attitudes are not derived only from doing and experiencing, and neither are they necessarily internally consistent with, or reflective of, any single identity, given the multiplicity of identities expressed in any one actor's life.

In Phulchi, natural resource use is politicized in a number of ways. Access is deeply conditional on social identity, and daliti, adivasi, and Bhumihar livelihoods all make different use of and affect environments differently. Furthermore, the understandings of the causes of perceived environmental problems reflect beliefs, prejudices, and emphases specific to the political stance of the actors. The environment is the site of political struggles, but it may not be the objective of those struggles in the manner environmentalists readily assume. The next section looks at self-perceptions, and those of others, in Phulchi, in order to see how identities are locally understood and how these connect with environmental relations and resource struggles.

DIKKU, ADIVASI, AND DALIT IDENTITIES AND STRUGGLES IN PHULCHI

The adivasi peoples of Phulchi are mostly Santal living slightly separately across the river and situated on the high tanr land, with a minority of Kohls who live in two small hamlets on the margins of the Bhumihar residential area and are in servitude to the Bhumihars. In our research, Kohls represented themselves as different from Santals, in being "more civilized" and not "jungle people" like Santals. Santals, however, did not necessarily see themselves this way; thus, for example, when speaking of local history, Santals emphasized a rosy past based not on forest livelihoods but on plentiful employment in the collieries that are now closed. Stereotypes of tribal identities found little resonance, and clearly the devaluation of "jungle" is not only found among Bhumihars but exists in uneasy tension with contradictory representations of closeness to nature by Jharkhandist politicians, and cultural entitlements to forests by Phulchi adivasis.

Dalits see themselves as the least advantaged group in the village, as the primary victims of Bhumihar exploitation, as dependent on adivasi political protection, but apprehensive of involvement with Lalkhandi activists. Their move to squat in the forest reserve, on the other side of the river from the Bhumihars, began a process of disengagement. Before this, some did not even have their own homes but stayed in or near the cattle sheds, and all members of their households were permanently "on call" to labor for the Bhumihar households (Bandyopadhyay 1983, 18). For dalits, the Jharkhand movement has emboldened them to resist and limit this exploitation, and while they continue to work for the Bhumihars, it is as casual laborers. One such man, for example, explained that even though they were not as able as the Santals to resist exploitation by the Bhumihars, they have gained freedoms in that they are not "compelled" to till Bhumihar lands as they used to be and can work in Giridih or elsewhere as they please. The exploitation and hypocrisy of Bhumihars is the subject of constant commentary. Women remark sarcastically on how pollution rules are disregarded by Bhumihars when there is hard work to be done by dalits for Bhumihars, but invoked over drinking water and cooked food when Bhumihars seek to humiliate them and assert their superiority. Dalit women take a quiet pleasure in the fact that when Bhumihar households need farm labor, Bhumihar men or women have to walk the short but symbolic distance to the dalits' hamlet and pass from house to house seeking (sometimes pleading) for laborers. However, at the same time, dalit sanskritizing influences are also evident in patterns of work of the relatively more prosperous dalit households and the trend toward dowry.

Bhumihars live in what is presented as the center of Phulchi, where houses are very large, close together, interspersed with walled barhi fields, and set around a tank, with a school at one side and a temple at the other. They always describe themselves as landowners and specialists in agriculture, and they deny any past or current involvement in mining and downplay any nonagricultural livelihood activities. Even those with little land take great pride in their caste identity as serious farmers. Adivasis and dalits regularly described the Bhumihars as obsessed with the control of "land and women." What might this persistent coupling of "land and women" mean; how are they linked, given the explicit denial of women's property rights, of their labor-based management and entitlements, and of their involvement with forests? In particular, how does the control of land involve the control of women? We return to these questions in the following section, but first we give an account of the resource struggles between groups.

Among the Bhumihar community, collection of firewood from the forests is most commonly done with bullock carts, by household males, and often with the help of hired labor. Trips are relatively infrequent for those who can transport large amounts but become more frequent for those who have to head load their fuel. Adivasis are attempting to stop the Bhumihars from using bullock carts to collect wood on the grounds that they are overusing the forest resource, and so some Bhumihar families are now head loading wood and having to make many more frequent trips as a consequence. The use of the forest is one area of conflict between adivasis and Bhumihars in which the former are bringing extensive pressure to bear to restrict Bhumihar access. The forest area has unmarked but informally recognized divisions into territories set aside for each of the three groups—dalits, adivasis, and Bhumihars—to use. During our fieldwork, a Bhumihar's house was surrounded by armed adivasis threatening violence in retribution for his being seen cart loading wood out of the dalit forest area. This is a new development, contrasting with the long-standing collusion of the forest guard (a Bhumihar himself) who turns a blind eye to Bhumihars cutting trees, according to the dalits. The solidarity of the adivasi with the dalit community is a consistent element of the changing patterns of power in Phulchi village.

The forest has been a key resource for the livestock elements of local livelihoods, especially of Bhumihars who herd stock to graze in the tanr and forest areas during the *kharif,* and also stall feed with forest branches. Stall feeding is practiced by Bhumihars and adivasis, although the Bhumihars do this for a larger proportion of the year and have larger numbers of stock and thus make greater demands on the forest. Cartloads of green branches with leaves are taken daily to Bhumihar households. This practice is against Forest Department rules, which allow collection of dry wood for domestic use but forbid the commercial use of the forest or the cutting of live trees. Adivasis are also implicated in the cutting of the forest for animal fodder, yet they are increasingly protesting, through direct action, against the level of Bhumihar fodder extraction. When Bhumihars express concern at the state of the forest, this is clearly linked to their experience of fodder as a constraint to livestock production. They say that livestock numbers are dropping because of the current problems with summer feed, "fewer leaves on the trees now," and the forest being "not so deep." Their sense of environmental scarcity is both grounded in livelihood experiences and expressive of the growing local *political* limits on their fodder extraction.

Struggles over land are perhaps the most important form of resource dispute in Phulchi, and they have deep and bitter roots. An adivasi man had sold a large area of land to Bhumihars who had killed his father and then persecuted the man by constantly letting loose their cattle on his crops—both his paddy and his barhi field—until eventually he sold them all his land for only Rs 7,000. Another adivasi man living near a Bhumihar related how he had had the legal title to his house until two years previous, when he had been tricked out of it. The Bhumihar said that a well was to be dug near the house, for which he needed to show the titles. After the title papers were handed over, only two to three feet of soil were dug before the work stopped, and the man has never had his papers returned. The theme of violence recurred in the accounts given of how the Bhumihars had historically secured most of the land of Phulchi village, when people say that murders over "land and women" were common.

In the past, the Bhumihars have acquired land through purchase and seizure, they have controlled the forests, and their political and social domination has been near absolute. They no longer have that control and are experiencing a resource crunch less because the environment has become degraded than because their rights to it are changing under the regional political currents asserting the adivasi rights above those of dikku. Despite their encroachments, Bhumihars perceive a shortage of arable land. This is because their uses of the forests are being challenged, the changes in the dalit community are raising issues over common property livestock grazing resources that had been taken for granted, and the Forest Department is less willing to collude in Bhumihar encroachment than in the past.

In 1976 the Bihar government passed the Land Restoration Act to prevent the wholesale acquisition of tribal lands by Bhumihars, under which tribal lands cannot be sold to a nontribal, unless the latter has had twelve years of possession. One consequence has been the encroachment by Bhumihars into the forest reserve and the deliberate manipulation of the lengthy legal procedures by which courts take decades to process land cases, so that by the time the judgment was made, encroachers have become owners. Bhumihars said that they sought to slow down the courts with so many cases that the delays would become even longer. Thus they continue their land accumulation, based now, however, on bureaucratic manipulation rather than direct violence.

Encroachment on the forest reserve land is practiced by all groups. The dalits have taken over an area for their residential settlement that is legally Forest Department land leased by the Damodar Valley Corporation (DVC). Both these institutions have made unsuccessful attempts to persuade them to leave, but there is now some recognition of their legitimacy, and the DVC has an arrangement that rewards them for protecting the planted forest. But the permission they have been given is only verbal; they have no titles, and they still say that they are "occupying the land. When the government comes we have to run away for a few days and then we return." The continuing tension derives from the encroachment into forest reserve of the arable fields that the dalits are establishing, for they are becoming farmers in their own right, albeit on a small scale.

A salient feature of resource struggles in Phulchi is that women as acting subjects are marginal. Struggles for forest use are between men, even where, as in the case of adivasi protests against Bhumihar cart loading, the issue of fuelwood provision is women's conjugal responsibility. Women also generally do not exercise their rights to land, although there are occasional attempts at this. A group of three Bhumihar sisters (without a brother) tried to challenge their uncles who had taken over their father's land after the death of first their father and then their mother, but the uncles refused to let the sisters have the land until they repaid the Rs 20,000 that the uncles claimed to have spent on their mother's funeral. Another widow had owned a large area of land but sold most of it, as she found it difficult to manage, and a third had willed her land to her only daughter and contracted a matrilocal marriage for her in Phulchi.

The dalit communities have little land for women to struggle over, but adivasi groups do recognize some rights of women to land. After a divorce or widowhood, a woman can remarry, whether or not she has children, and when the children are grown up, they may claim the land of their fathers. But land in a widow's name is not granted, and a widow who does not remarry is only allowed to use her husband's land to feed herself. This situation has led to many witchcraft allegations and attacks and murders of widows using land from which male relatives wish to see them dislodged. Kelkar and Nathan (1991), surveying women's land rights in the Jharkhand, suggest that it is precisely where widows have strongest rights to land use that witchcraft abuses are most severe.

We have argued here that ethnicity and gender sharply differentiate the environmental relations of Phulchi's residents within a changing context of political and gender ideologies. The Jharkhand movement has allowed adivasi men to reclaim resources on the basis of their tribal identity, and to offer a cultural patronage to low-caste groups in the village. This has deeply eroded Bhumihar confidence in their right to resources, even if it has yet to dramatically alter landholding patterns. However, village discourses of Jharkhandi identity are oriented toward forest and land entitlements, not toward harmony with nature or the absence of patriarchy, and are firmly grounded in local livelihood and agrarian relations. Women's limited struggles for land are more characteristic of Bhumihars than adivasis or dalits, most of whom have little land to struggle over. Bina Agarwal (1994) suggests that wealthier women are less likely to try to obtain land rights, since they have more to lose in the familial acrimony that follows land claims by women, but on the other hand, women of poorer groups are likely to be effectively landless. This much is clear, that diverse political, ethnic, and gender identities legitimate some resource claims and deny others, and that "nature as idea" has been effectively deployed in adivasi and dalit, but not in gender-based, struggles.

However, identity discourses that suggest that dikku/men exploit nature and adivasi/women revere it need to be examined in the light of manifest environmental actions and outcomes, in actually existing agrarian livelihoods. It is to this, especially the gendered expectations of the dualism, that we now turn, because as we have pointed out, nature exists not only as idea but as managed livelihoods, and the concept of agrarian environments developed by Agrawal and Sivaramakrishnan directs attention to the integrated lived experience of survival and change for different social groups in Phulchi.

The concept of agrarian environments is usefully compatible with the integration of production and reproduction that feminist analysts have insisted on since the 1970s. Livelihoods require reproductive labor. The domestic work of women and men in direct provisioning place people, especially women, in sometimes intense daily interactions at the interface of the agrarian and the environmental, which are lived as unified. The brief accounts that follow show the livelihood positioning of different groups of women, and the ways in which they connect, and disconnect, men and women from direct interaction with nature. They show the gender ideological context of agrarian environments and how these too both divide and define ethnicity, as well as implicitly unite male gender interests.

The three social groups of Phulchi have significantly different livelihoods and gender relations. Bhumihar livelihoods rely heavily on agriculture and depend on cultivation of patrilineally inherited land, the average being 4.7 acres in Phulchi village, excluding holdings in other villages. Bhumihars nevertheless express a sense of land scarcity, pointing to the rising price of land as evidence, and they actively pursue land accumulation strategies, by purchase and marriage transactions, as well as intensification initiatives.[8] Bhumihar farmers employ large numbers of wage laborers, use fertilizers and compost to maintain soil fertility, and sell surplus paddy in most years. Most Bhumihar men do not allow their wives to do farmwork on their own lands, other than in walled barhi plots, or those of others. Such men are also firmly of the opinion that they work harder than women, saying that work like soil cutting and plowing is heavy work that women cannot do.[9] The involvement of women in discussions about farming, such as crop choices, is minimal, with the exception of barhi land. Unsurprisingly Bhumihar men denied that women had any knowledge of crops, trees, soils, or natural resources and insisted that they, the men, were the knowledgeable ones. In one farmer's words, "The men go here and there and see everything, they have to manage the farming so they have to learn from others, and they sit and talk about farming, discussing everything. Women do not do this, they only do housework." For their part, women concurred in the devaluation of their effort, their skills, and their agricultural knowledges.

The arable-land-based livelihoods of Bhumihars are part of a patriarchal complex in which women are strictly controlled, their numbers regulated, education denied girls, and early marriage with heavy dowry practiced. Of the twenty living children born to the seven Bhumihar women in the intensively studied households, only four were girls. Stories of female infanticide were discussed, and sex-selective abortion, although hard to substantiate, is so widely known about (in terms of the location of clinics, the costs, and procedures) among Bhumihar women that it seemed likely to be a feature of reproductive strategies of a number of Bhumihar households. No other social group had the same distorted sex ratio among children or were as familiar with sex-selective abortion.

Hypergamy severs the support of natal families for married women. Bhumihar women cannot divorce, and violence against wives (by husbands and mothers-in-law) is socially sanctioned. Wives lack autonomous incomes or recognition of their productive work and are generally unable to acquire either the use or ownership of land. Seclusion practices limit their mobility. Reproductive strategies aim, usually effectively, at few and almost exclusively male children to ensure consolidated landholdings, which are used to attract very high dowries for sons. Although Bhumihar women live in more prosperous households, they also exist within a context of more oppressive gender relations and a more profoundly alienated relationship to natural resources. Male obsessions with land implicate the control of women, both in terms of the numbers of women (in sex-selective reproduction to limit the numbers of daughters whose dowry needs can threaten land assets) and in terms of their tractability as wives absorbed into and serving the lineage.[10]

In such circumstances, not only are Bhumihar women removed from direct relations with nature, but land is the root of much that is oppressive in their lives. Their opportunities for agency and resistance lie in the character of relations with men, and indeed Bhumihar women's gender struggles are mainly about access to men as resources.

Bhumihar women are members of the social group most overtly committed to agriculture, and to "sustainable" rural futures, yet they personally have limited direct labor, property, or management-based roles in these livelihoods. Furthermore, Bhumihar women may well be victims of their own caste identity, but they simultaneously subscribe to Bhumihar domination, support castism, and do not express solidarity with low-caste or tribal women. Generalized statements about women as active agriculturalists and as environmental managers, carers, and defenders have little application to their lives.

In adivasi livelihoods, arable landholdings are either absent or very small, ranging to a maximum of a few acres, and almost all adivasi households stressed laboring in their accounts of their livelihoods, some to the exclusion of all else. Laboring work is done by both men and women, and both prefer to work locally rather than go "outside," but almost all men do go to Giridih for daily work in the dry season, and a number migrate further afield for longer periods and send remittances home.[11] There are no gender wage differentials, and local demand for wage laborers is seasonally high.

In own-account farming by adivasis, little hired labor is employed, and there are no generalized seclusion practices, so that family women find themselves heavily involved in household farm labor, but on small acres of land. Most animals are small stock, herded to feed and tended by family women. The area of greatest management by women is barhi land, where labor and decision making are controlled by women, and on garha land there is least involvement by women. The value of women's work is acknowledged, and tribal men (unlike other social groups) usually gave gifts, saris and bangles, to their wives after harvest in recognition of their farmwork contributions.

Adivasi men, however, denied that women had a greater or special knowledge of trees, saying that men know more species and talk about trees among themselves and also with the DVC and the Forest Department.[12] Further, although women are heavily involved with farming, they are denied any recognized expertise in farming.[13] This is not paradoxical locally, however, because women's "farming" is as laborers, not managers. Men, too, who had worked for decades in laboring on every possible stage of crop production often represented themselves as "not knowing" how to farm because of the connotations of skill and knowledge in management and of unskilled manual work in farm laboring. Conversely, Bhumihar men do relatively little farm labor, mainly supervising farmworkers, but their expertise is seen to reside in qualities such as their innovativeness. Models of knowledge systems that dualistically oppose *technē* and *epistēmē,* associated with Hindu and Western cultures respectively (Marglin 1990, 277), and that represent the hierarchy within *technē* as age based and therefore not inherently inegalitarian, fail to recognize *epistēmē* in "traditional" societies, and the gendered exclusions in both knowledge systems.[14]

Clearly, male adivasi perceptions of women's farmwork are complex. The wealthier tribal families showed a tendency toward withdrawal of women from farmwork, first a refusal to work as hired laborers for other farmers, and second a withdrawal from family farmwork, too. Tribal men held the view, however, like Bhumihar men, that men work much harder than women. Although domestic work was seen as important, one man saying, "I could not do it!" it was also seen as light, another adding dismissively that women's work is "only cooking, bringing water from the well." This was although tribal women are more involved with wood collection, certainly heavy work, than either of the other two groups and have much heavier work in water collection than dalit or Bhumihar women, since the distances to water are considerably greater, and the terrain steep. Some men do help their wives with wood collection, but it was largely seen as a female responsibility. In summary, adivasi women in Phulchi do not have significant direct land rights, and they are heavily involved in family farmwork, but on holdings reduced by Bhumihar incursions. They are largely farmworkers rather than farm managers, considered agriculturally ignorant by men and not directly involved in resource struggles despite their load of reproductive work.

From a position of almost no land ownership some decades ago, dalit livelihoods are becoming more land based over time as a result of disengagement from Bhumihar labor contracts, forest encroachment, allocation by the government, and a little purchasing of land. Moving out to their new hamlet ended many dalit labor contracts, and thereby their access to very limited use rights on Bhumihar land, but simultaneously opened up the possibility of expanding and securing new landholdings. Land expansions are oriented toward productive garha land, with wives playing a major role in land preparation during the *rabi,* when there is little local farmwork available. Although wage labor is the most significant element in the livelihoods of the nearly landless dalits, farming is of growing importance; few dalits have no own-account farming activities, and some see farming as more important than laboring. Many now raise goats and pigs both on a share system with Bhumihars and on their own account, explaining that when they lived in close proximity with the Bhumihars, they were unable to raise animals because of the persecution that resulted if their animals ever entered a Bhumihar compound.

Both men and women do wage-laboring work, and by all accounts there is no differential between local wages paid for farmwork by men or women, and little movement between villages for farm labor. As one dalit remarked, "Here there is so much work." However, some dalit women appear to be withdrawing from wage work to some extent and concentrating on barhi cultivation at certain seasonal points. Wages have risen on two occasions in recent memory. The first was in 1988 to 1989, when Sibu Soren encouraged protest by dalit and adivasi communities, and the second was in 1991 when all the (male) dalit laborers went to the village head and successfully demanded an increase in the daily rate from three to four *paila* of rice.

Domestic labor by women is the norm, but many men do help with wood collection, and dalits were the only group in which men mentioned taking over domestic work when their wives were particularly busy. Son preference is far from absent in dalit culture, but it is not linked with the dramatic imbalance in sex ratios found in the Bhumihar community. Conjugal contracts stress a joint responsibility for "earning," and divorce and remarriage are accepted in unhappy marriages.

Phulchi identities and livelihoods are divided in dramatic ways but also woven into complex and shifting patterns. People speak of deeply oppositional identities, which acquire meaning in relation to each other, as members of ethnic groups and of genders within those groups; but at the same time, we observed very interdependent livelihoods. There are transactions of knowledge, labor, and livestock within local livelihoods that bind, in addition to the struggles that divide. The most significant local agricultural change taking place is the intensification in resource use whereby the area irrigated is small but growing. Well irrigation of barhi fields is expanding, and most households grow some dry-season crops through the use of residual moisture.[15] Considerable energy is being invested in fencing and protecting barhi fields from livestock to enable dry-season cropping, often of high-value crops such as potato, based on the use of fertilizers and compost. Many farmers have tried to make soil conservation structures to prevent soil erosion; almost all have been planting fruit trees around their homesteads; and intercropping has spread widely. These practices are not confined to the wealthy, and even very marginal farmers adopt and adapt new ideas, albeit on a small scale.

Innovative practices in Phulchi clearly begin with the Bhumihar community, and both adivasis and dalits acknowledge that the changes to their farming practices generally come from observing successful experimentation by Bhumihars; and in this way, dry-season cropping, well irrigation, intercropping, and composting have spread to the smaller adivasis and dalit farmers. Dalit farmers, though distancing themselves in some ways, do continue their relationships with Bhumihars. For example, one man who has broken off a bonded relationship with a Bhumihar nevertheless now gets his seed from his former patron and says that he decides what to grow by watching what the Bhumihar grows, and his methods of experimentation, "observing for one year another variety in a separate plot." At the same time, adivasis remark with satisfaction that in the old days, they used to rely on the Bhumihars to lend them seed for their own cultivation, but nowadays they are able to buy seed from markets and to retain their seed from one year to the next. What we see in Phulchi is multiple levels of dependency as well as antagonism between social groups. Bhumihar men value farming identities most, have the greatest level of livelihood commitment to arable lands, and support local agricultural knowledge systems, but they simultaneously exercise a hegemony that exploits and controls dalit and adivasi. Furthermore, Phulchi men as a group share a belief in the necessity of controlling women, and male gender interests are in some ways a bridgehead across ethnic divisions.

This section has described the divisions of labor and management by ethnic group and gender and has shown how diverse and specific are the day-to-day environmental relations of particular groups of people in terms of the work they do, the knowledges they have or are perceived to have, the responsibilities they carry, and their perceptions of others. The processes of gender exclusion and identification in Phulchi livelihoods are both material and ideological; divisions of labor place women as workers and construct them as unskilled, and local epistemologies of what counts as agroecological knowledge and expertise elaborate and reflect this construction. Knowledge is seen as a product of public discourse between men rather than private encounters with nature in the course of a working life, and by this criteria, too, expertise is gendered. These exclusions operate for all women in Phulchi, but they are possibly less significant for Bhumihar women, who are excluded anyway from farm and forest work, just as, despite the wide variation in livelihoods, men of all groups share views on women's knowledges.

CONCLUSIONS

The Jharkhand movement has generated anxiety among Phulchi's Bhumihars, both through the changing character of the local state and the sense of legitimacy surrounding the claims of the admittedly reformist elements and through the fear of direct action by Lalkhandists. It has also created the conditions under which adivasis have been able to protest against levels of forest use by Bhumihars, and dalits have been able to move to much greater spatial and social autonomy from Bhumihar domination. However, these shifts are driven not by adivasi concepts of "harmony with nature" but by changing abilities to compete for local resources, and it is a livelihood perspective that reveals this, as well as the complexity of local interdependence. The ability to compete for local natural resources is also seen in Phulchi to be critically dependent on the regional political economy, in which ethnic identities are imagined and ideas of what adivasis are come to have profound effects on material livelihoods.

What also emerges here is the significance of gender ideologies and ideas in mediating the environmental relations of women as well as men. Bhumihar women relate to land indirectly and through marriage circuits in which son preference and the control of women are powerfully inscribed. Despite their more direct relations to land and forests, adivasi women's environmental relations remain framed by their gendered subject positions. Ideas and identities are not straightforwardly derived from material livelihoods and property relations but have a life of their own, drawing in actors and interests far beyond the boundaries of the village. Yet equally discourses of gender and ethnicity have personal meaning only in the context of actually existing livelihoods and social relations; Jharkhandist ideas are drawn on locally insofar as they enable the redirection of resource struggles by men or struggles against sexual abuse of women. The relative autonomy of identity discourses is also seen in perceptions of local knowledge in which livelihood practices are not seen to amount to "knowledge." Being a woman resource user or a male or female farmworker is irrelevant in local conceptions of knowledge that is considered to reside in those engaged in public discourses, and to have managerial control of resources rather than an embodied and practical engagement with nature.

Nature as idea can be distinguished from nature managed through livelihood practices at the same time as it is conceived of as being variously connected to such practices. This is no different from the claim that gender ideologies can be distinct from the actual lives of men and women even as they are connected to lives. Once these distinctions and relationships are conceived of as coexisting simultaneously, then debates about social constructionism versus realism in environmental thought can fruitfully be brought together.

NOTES

1 Terminology is always fraught in referring to "ethnic" groups. The Indian convention of Scheduled Castes and Scheduled Tribes labels and imprisons "the poor" in a kind of inverted bureaucratic castism and is also imprecise. On the whole, we use the terms *adivasi* to refer to the tribal groups of the area, *dalit* to refer to the local Scheduled Castes, who are mostly Turi, because these are the terms most frequently used in self-description. We use either *dikku,* a pejorative term used by the former two groups to refer to the Bhumihar (agricultural caste Hindus), or simply their caste name.

2 Gaard's arguments all assume a subject living in an industrialized Western environment who needs to experience weather, "the simple act of walking," distant horizons for the eyes, the healing effects of the magnetism of sacred places, natural and bodily smells, and calming sounds (1997, 16–23). Nowhere does she recognize that providing these experiences for priviledged Westerners has social justice implications both within the West and globally, when large swaths of the south are deployed toward the tourism that meets these "needs."

3 During late 1993 and 1994 the authors researched the village in this study with financial support from the Overseas Development Administration's Population and Environment Research Programme. We gratefully acknowledge the support of the Indian Statistical Institute in Calcutta.

4 The Jharkhand area of south Bihar, hilly, forested, rich in minerals, and home to the great majority of the tribal population of the state, together with adjacent districts in West Bengal, Orissa, and Madhya Pradesh, forms Greater Jharkhand. This area has been articulating demands for separate status since 1928, most recently, since 1972, by the Jharkhand Party. Land alienation from the indigenous peoples has taken place for centuries, and the opposition between dikku and adivasi was well established by the colonial period. But the British legal and administrative changes that strengthened the outsiders at the expense of tribals were the background to the anticolonial and ethno-nationalist peasant movements for which the area is well known. Industrialization and urbanization in the region attracted immigrants and were not seen to benefit local people, since minerals and timber are exported, forests

have been protected and adivasi access restricted, electricity generated goes to the northern plains, and development initiatives by the state are both paltry and corrupt. Epic tales of the corruption of state officials are a major part of oral culture in Phulchi.

5 The Giridih area has a considerable mixture of tribal and caste populations from which there is a southward movement to the more thoroughly tribal areas, for example, around Ranchi.

6 That there are only a few government-subsidized wells and many privately built Bhumihar wells was not seen as relevant, nor did it modify their resentment.

7 Dualistic terms such as *insider* and *outsider* are crude shorthand and not intended to imply a watertight distinction, but simply to indicate broad positionality.

8 For example, Mahinder Roy bought ajan land six years ago and sold tanr land four years ago, saying that he is not interested in land that cannot be irrigated.

9 Most domestic work falls to women, such as collecting water (usually from compound wells), feeding livestock, cleaning, and milking, with the exception of washing buffalo, collecting fodder outside the homestead, and collecting fuelwood, which are men's work. Women also have to cook meals for large numbers of laborers during peak seasons.

10 The subversions of this tractability by Bhumihar women are many—but this is another story.

11 Part of the preference for local work has to do with the physical effort involved in wage work in construction and nonagricultural activities, which is much higher than farmwork, as well as the exhaustion related to the long traveling time.

12 However, we encountered individual women who were much more knowledgeable about soil differences than their husbands.

13 An adivasi man, when asked about whether he consults his wife about farming decisions, replied, "Why should I? She does not know anything!"

14 Marglin does footnote that hierarchies of sex exist within *technē*, but he excuses these on the grounds that "traditional societies typically allocate separate *technai* to men and women, so that in so far as power relationships between sexes are related to knowledge they reflect rules of power *between* rather than *within* knowledge systems" (1990, 235). Both *technē* and *epistēmē* separate, confine, and devalue women's knowledge, in ways that qualify an uncritical enthusiasm for the "local" and suggest a rethinking of *technē, epistēmē,* and disadvantage.

15 Planting a crop at the end of the wet season that matures on the stored water in the soil to produce a dry-season crop.

Shubhra Gururani

Regimes of Control, Strategies of Access: Politics of Forest Use in the Uttarakhand Himalaya, India

On a late December afternoon sitting on a rooftop, Amma, now in her mid-eighties, said to me, "Where are the forests? There are no trees left. There is nothing out there now. Nothing! It was not like this before. This was a very thick forest, full of oak. It used to be so thick that we could not see the other side of the mountain. There were so many wild animals. We used to be scared to go into the forest. It was not hard to find firewood or fodder. . . . I would put my young son to sleep and run to the forest and bring back enough fuelwood and fodder even before he turned on his side. It used to hardly take any time to collect firewood and fodder. . . . These days it is like a war *(yudh)*, women spend whole days in the forest. Forest guard sits in the forest like a wolf *(siyar)* waiting to catch the villagers, government *(sarkar)* has [imposed] rules/barriers *(rok)* on the forest. . . . Now women are in the forest from sunrise to sundown, and it is still not enough. . . . Life has become so much harder."

The story Amma tells of rich forests is told by many in the Gorung Valley in Pithoragarh district in the Uttarakhand Himalaya in India.[1] Those who have seen a good part of the century recall better times and lament the state of degraded forests and declining productivity of the land. Embedded in the memories of wooded forest and abundant game are the images of forest guards patrolling the forest. Ever since the villagers can remember, they recall the presence of a forest guard and recount their many encounters with him. Most villagers do not remember a time when the entire forest "belonged" to them and clearly recall the division of the forest into reserved and village *(panchayat)* forest. Some have a fragmented recollection of rules and restrictions that have forbidden them to cut leaves and branches for almost a century now.

Even today the reserved and panchayat forests in Uttarakhand are clearly distinguishable. The majority of villagers, including children, identify the exact boundaries of the two forests facing their village and enumerate the many restrictions that define the boundaries of use and access. The knowledge of forest boundaries and some awareness of the rules and regulations of use, however, have not stopped the villagers from entering the forest over the past several decades. They rely on the long tradition of collecting fuel and fodder from forests designated as reserved and devise strategies to claim a share of the forest. The complex and creative mechanisms and relations that permit use, access, and withdrawal of forest products in a Uttarakhand village present an interesting case to rethink the two concepts that have underlined the debates and discussions of management, governance, and control of resources — the concepts of property and community.

In this essay, by uncovering the many layers and matrices in which both property and community are located, I want to focus first on property rights and show that the rules and codes of property established by the Forest Department are one set of (legal) limits posed by property rights regimes. There also exist deeply embedded cultural, historical, and economic claims, although de facto, that modify property rights and help us understand how people use their resources and lay a claim on them. Looking at property not as a thing but as a relationship (Rose 1994, 5), my attempt here is to show the socially constructed and historically sedimented nature of property that allows rules and codes of property to be redefined and reinterpreted beyond their narrow legal confines. To emphasize the complex and multistranded construction of property and community, this chapter focuses through an anthropological lens on the subtleties, practice, memory, landscape, and local narratives that describe structures of assertion and narratives of legitimacy to resources otherwise deemed inaccessible by the code of law.

In the first section, I briefly discuss the debates that surround the discourse of property and argue for an understanding of property and its rules as a set of power relations and as practice. I wish to argue that practices of property are produced within the context of regional political economy, local history, and landscape. Although the distinctions between private, common, and state property exist on paper, they are ambiguous in the realm of everyday practice. By focusing on these practices of claim, I want to demonstrate how the competing claims and overlapping boundaries of moral and legal property, rights and livelihood, and rules and practice work to redefine and reinterpret property in the context of cultural politics.

In the second section, to explore the specific representation of community that underscores the discussions of property, I highlight the gendered and caste- and class-based nature of resource use. I show how the villagers constantly redefine and renegotiate the rules and codes of property along caste, class, gender, and age lines, demonstrating the conflicting interests and motivations that coalesce or fragment communities in different periods over different resources. A nuanced approach to struggles over resources shows that the struggles and claims over resources are fundamentally about power and control and are aimed at asserting dominance of one regime of property over another.

For my analysis, I draw examples from both reserved and state-owned forests and village commons in Kumaon Himalaya. Although a focus on the village commons merits more detailed discussion and has been the subject of many studies,[2] a discussion of practices in the reserves presents an equally interesting picture from the perspective of the property literature. The reserves, owned and managed by the state-run Forest Department, constitute more than two-thirds of the total forested area in Uttarakhand and supply a large proportion of fuel and fodder to the villagers. Villagers do not have formal rights of harvest of forest products, but they continue to cut leaves and branches on a regular basis. The everyday discourse and practice of claim, access, and withdrawal in the reserves provide a useful analytical entry into the discussion of resource governance and control.

Importantly, an ethnography of the reserves presents a critical counterpoint to the constitution of rural resource users as selfless communitarians. Unlike village-owned forests (*ban panchayat* forests), which are shared by most or all members of one or two villages who through village councils design the rules of access and withdrawal *(res communis),* reserved forests are shared by a large number of different caste villages, disaggregating the simplistic notion of a homogeneous community and revealing the multi-stranded relations of conflict and consensus that inform community relations.

Finally, the reserves are a case where the issues of authority, monitoring, and responsibility are encountered through the politics of gatekeeping and boundary making. The everyday confrontation that ensues between the forest guard and the villagers presents a case in point to explore the multiple levels at which the state officials interact with the people and how the villagers contest the officials' authority. The elaboration of the contest between the forest guard and the villagers urges us to explore the various social and political alliances that allow and facilitate the nonobservance of the rights of the Forest Department and create room for redefining and reinterpreting property and its rules. Such an ethnographic study allows us to rethink property and community as the mainstay of common property literature and show how state ideologies of conservation and degradation are produced and how they are resisted, negotiated, and redefined.

RIGHTS, DUTIES, AND RULES OF PROPERTY

It is generally agreed that property implies "a system of relations between individuals. . . . it involves rights, duties, powers, privileges, forbearance, etc., of certain kinds" (Hallowell 1943, quoted in Feder and Feeny 1991).[3] Property is treated as a bounded set of rights that allow individuals (private) or groups (common or corporate) to press their claim on a current or a future benefit stream, a resource, that is enforced by society or state, by custom or convention or law (Macpherson 1978, 3). Such a focus, which regards property as a bundle of rights and duties, tends to overlook the many structures and relations of control and authority that shape all social networks and consequently property.

Rights over property are usually distinct and define the limits of resource use and management. Rights distinguish between access and withdrawal of resources at the operational level and management, exclusion, and alienation at the collective-choice level. Access rights give the right of entry into a defined physical property, whereas the rights of withdrawal are the rights to obtain the products of a resource (see Schlager and Ostrom 1992). In the reserved forests of Uttarakhand, all villagers, according to this definition, have the right to enter the reserves, or they have access rights, but the withdrawal rights are limited to dry and fallen leaves and wood. The villagers, however, do not have any collective-choice rights that give them the right to regulate internal-use patterns (management rights), or the right to determine who will have the right of access (exclusion right), or the right to sell or lease management or exclusion rights (alienation rights) (252). The Forest Department retains all operational as well as collective-choice rights. Even though the villagers are unable to formally change or contest the collective-choice rights of management, exclusion, or alienation held by the state, they regularly attempt to redefine and transform the operational-level rights. Despite a clear understanding of their legal rights, villagers regularly violate them and withdraw fuel and fodder from the reserves.

State property, although it is not open access and clearly belongs to the state, is treated as if it is open access. De jure rights are established by the state, and the villagers, in this case, exercise de facto rights on state property. Schlager and Ostrom (1992) argue that "a conglomeration of de jure and de facto property rights may exist which overlap, complement, or even conflict with one another" (254). The reason why de facto rights coexist with de jure rights is linked to a key distinction, pointed out by Ribot (1997), between *rights* and *ability,* which provides an interesting point to interrogate the formal and informal implications of rights in the realm of practice. Whereas rights present the prescriptive side, ability is "broader than right, resting solely on the fact of demonstration without the need for any socially articulated approval" (Ribot 1997). Most often, those without formal rights exercise their ability to use property and overlook or break the rules. That people exercise their ability and break rules is not interesting in itself, but the cultural, moral, political, and historical bases that guide the ability to break rules are critical for our purpose here. That is, the ability to use and access forest suggests alternate spheres in which property is practiced and allows us to look at property as a multilayered set of power relations informed by the cultural, regional, and historical contexts.

The issue of rights is linked to rules. Rights over property are regulated by a set of rules that as forced prescriptions authorize the rights of the owner, proprietor, claimant, or authorized user either by custom or by law (Schlager and Ostrom 1992). Rights are seen as products of rules, and to be operative, rules demand acknowledgment and enforcement. The acknowledgment and enforcement of rules ties rights to duties, as acknowledgment precludes understanding of not only their rights but also those of others. Bromley writes, "Property is a right to a benefit stream that is only as secure as the *duty* of all the others to respect the conditions that protect that stream. . . . We can characterize these relations between two

174 *Shubhra Gururani*

(or more) individuals (or groups) by stating that one party has an interest that is protected by a *right* only when all others have a *duty*" (1989, 871). Rights and duties together are thus products of rules.

Although forest rules are central to the definition and implementation of rights and duties, they are effective only if they are recognized and observed by those who are expected to follow them. The nonrecognition or breaking of rules thus reinterprets the rights and duties in *informal* ways that are important pointers for us to conceptualize rules as negotiable and arbitrary. To explore the interface at which property gets redefined and modified, in the following section, I discuss rules of property in everyday practice.

PRACTICES OF PROPERTY

Rules as "euphemized form of legalism" become arenas where struggles over resources and over meaning are expressed (Li 1996; Berry 1988). Rules and lines of property are constantly contested, negotiated, and reproduced in ways that confront our understanding of rules as neutral templates of interaction. Rules are, however, linked to practice, and as Taylor writes, "practice is a continual 'interpretation' and reinterpretation of what the rule really means. . . . They [rule and practice] renew it and at the same time alter it. Their relation is thus reciprocal. . . . *In fact, what this reciprocity shows is that rule lies essentially in practice*" (1993, 57; italics mine). Practices of forest use and access are highly complex and ambiguous and often "express a social practice that in fact obeys quite different principles" (Bourdieu 1977, 19). Although rules as "structuring structures" shape practice, they are also shaped by social practice. Practice thus is a product of structures that are generated by complex processes of manipulation, mediation, and individual and group dispositions and often diverge from the prescriptions laid out by rule (Sivaramakrishnan 1997) and in fact become rules of sorts.

The range of mechanisms and schema devised by "nonauthorized" users in Uttarakhand describes the contradictions that characterize the apparatus of rule making and implementation (Agrawal 1994c). Rules are constantly contested, resisted, redefined, and reinterpreted in ways that not only highlight the complex interface between rules and practice but show that rules are subject to multistranded power relations shaped along the lines of caste, class, and gender. As Sivaramakrishnan points out, rules are vulnerable to modification resulting from conflict and cooperation between the state and society: "Political discourses, including laws, rules, codes, and official manuals, create both unifying *and* fragmenting possibilities by establishing new fields of political struggle. . . . This is what allows social actors as humble as graziers and woodcutters to leave their impress on political discourse through the very structures of bureaucracy and its internal contradictions" (1997, 79). The issue of conflict within the government influences the formation and implementation of rules and is an interesting one and well elaborated by Rangan (1997) and Saberwal (this volume) but is beyond the scope of this paper. It will, however, suffice here to say that the long history of tension between the Forest and Revenue Departments has left ambiguous spaces in rules that some villagers recognize and manipulate.[4] In this chapter, I focus only on the local politics and demonstrate how the relations of class, caste, and gender not only determine practices of forest use and access but also color the ways in which villagers interact with those who implement and enforce the rules.

To discuss the socially constituted nature of property and its rules, I find the focus on everyday discourse and practice through which people use and access forest useful. Following Tania Li, I use a "micropolitical economy" approach, which emphasizes "human agency or *praxis,* focusing on the creative ways in which cultural ideas are adapted to meet new conditions, and culturally informed practices, in turn, structure daily life and shape and reshape institutions at various levels" (1996, 509). Such an ethnographically thick approach will describe not only how rules are redefined and negotiated but also the mechanism through which the rule makers are confronted in subtle but not uncertain terms.

LANDSCAPE OF PROPERTY

Village boundaries in Kumaon were demarcated and fixed from the time before pre-Gurkha rulers and were recorded in the government land records with the first British settlement by G. W. Traill in 1823 and have since been popularly known as "assi sal bandobast," corresponding with the Samvat 1880 in the Hindu calendar. According to this revenue settlement, villagers maintained certain rights of grazing and fuel in neighboring forests (Ramachandra Guha 1989b, 32). With the establishment of the Forest

Department in 1878, the redefinition of forest into different classes resulted in the sovereign's *legal* claim over the forest. Gradually, guided primarily by commercial interest in resin and pine for railway sleepers, large tracts of forests were brought under state control, and several species of trees were demarcated as reserved (Pant 1922, 14–16). The customary rights of the villagers were abolished, and they were given certain concessions that were notified by the durbar (40). For example, villagers were allotted fixed amounts of building timber, as well as the privilege of collecting dry fallen wood and cut grass in specified areas. Free grazing was also allowed in the second- and third-class forests within a five-mile radius of each village (40). Apart from these privileges, any fresh clearing, felling, girdling, lopping, tapping, or grazing of cattle in the areas demarcated as "reserved forests" was strictly prohibited and considered a forest offense (Forest Manual 1936). The villagers, as Pant (1922) and Ramachandra Guha (1989b) report, resisted the growing control of the state over their forests, which affected their subsistence practices directly. In response to these uprisings, the rules of access and withdrawal underwent some revisions over the years, but access to reserved forests continues to be restricted, and withdrawal of fuel and fodder is limited.

Currently, the entire supply of fuelwood and fodder for the village I call Bankhali, comes from the reserved forest.[5] The 247 hectares of reserved forest facing Bankhali are surrounded by twenty-eight villages and are guarded by a forest guard who belongs to the Brahmin caste. The forest guard is locally called *patraul,* from the word "patrol," and sometimes the "bull" or the "wolf"; and the common refrain among the women is that they will be attacked either by a man-eater or by a money-eater (patraul). For five months in a year (November to March), Bankhali women visit the forest at least twice, and sometimes thrice, a day. They walk two to three hours each way, covering three to five miles through pine and oak forests, and each brings back heavy loads of mainly oak wood or leaves, weighing not less than thirty kilos on each trip.

The guard, according to the rules of access and withdrawal, is expected to catch any offender found cutting fresh leaves or branches or trees in the reserved forest and to take away their implements and also fine them. Women are allowed to collect only dried fallen wood. To avoid the guard, women time their trips to the reserved forest carefully and go to the forest only after they have made sure that the forest guard has gone past their patch of the forest. The activities of the forest guard are persistently monitored, and even men and women who do not go to the forest keep track of the forest guard. Discussions about the forest guard constitute a significant part of conversation among Bankhali men and women.

All the women recognize that the forest "belongs to" the government and that there are strict restrictions regarding the use of most forest products. They are aware that they are not allowed to cut fresh leaves or branches from the reserved forest and that it is an offense to cut down branches in the eyes of the Forest Department, but they nonetheless insist that the demands of subsistence force them to go against the rules of the department. One of the women said,

> We know we steal wood from the forest. The forest is *sarkari* (government) and we are not allowed to go there but we have no other way of getting fuelwood or fodder. We have to steal wood to live. If we do not steal we cannot live in this cold, we cannot cook, our cattle cannot live, we will not have any manure. . . . Why should the sarkar punish us? We never ever sell wood or any other products of the forest. We steal for our families.

That women break rules and steal from the forest is not an unknown scenario and has been widely discussed in the common property literature, but what is of relevance here are the many interpretations offered by different members of the community to explain stealing. Pursuing the famous phrase and logic of property as theft (see Rose 1994, 2), then maybe we can play with the possibility of its obverse, namely, theft as a claim to property. Acknowledging the limits of this argument, I nonetheless think it is interesting to look at stealing and villagers' responses to it, not only as resistance but as enunciation of informal spheres of claim and entitlement that are determined by social, cultural, regional, and historical conditions that have persisted for a long period.

Given that the majority of villagers in the Uttarakhand Himalaya are largely subsistence farmers, their extreme dependence on the forest is connected to the demands for fuelwood and fodder. The particular climatology of the hills and their rugged terrain do not allow for irrigated farming. As in most other Himalayan regions, crops in Uttarakhand are primarily rain fed. The lack of any well-developed irrigation system leaves the villagers at the mercy of rain. Because rain-fed soil is low in humus content, it is necessary to raise the humus content, which is done by using organic fertilizer. Mainly cow dung is used to increase the moisture-retaining capacity of the soil. Because cattle dung provides the primary source of organic fertilizer for rain-fed farming, women say that they have no choice but to steal grass and oak leaves to generate organic fertilizer. There are hardly any alternate means of fodder, and even when there is some fodder available, it barely reaches a quarter of them.

Similarly, the women argue that in the absence of any alternate means of fuel, they are forced to ignore the rules established by the state and monitored by the guard. They claim that the unavailability and high costs of cooking gas cylinders and kerosene oil leave no other way to meet their subsistence needs but to rely on the reserves. According to them, even if their husbands and sons send remittances home, it is the women's contribution from the forest and farm that actually keeps the hearth going.[6] They elaborate their long hours of hard work, show their bruised arms and legs, complain about their health, and say that the life of a mountain woman is the worst of the lot. They claim that in the absence of any other means of fuel or fodder, as good wives and mothers, they are forced to go to the forest against all odds. They say that they trick the forest guard, lie to him, and sometimes threaten him for the sake of their children and husbands. By upholding the ideology of the hearth and demonstrating their dependence on the reserves, women make moral claims of subsistence on the forest. The boundaries of the reserves are erased by the logic of hearth and farm and transform the reserves into a landscape of moral discourse and practice. In the absence of alternate means of fuel and fodder, women press their claims of livelihood and dependence, which allow inscription of informal claims and entitlement on the reserves and challenge state ideologies of conservation and protection.

It is important to point out here that women do not make a claim on the entire forest per se. I never heard anyone claiming the forest, but many insisted that they should be allowed to lop branches and collect grass for their cattle. Nor did the villagers seek forest products for commercial use.[7] Women claim withdrawal rights over fuelwood and fodder primarily for subsistence. They say they have always collected fuelwood and fodder from the forest and do not see any other way to support themselves (see also Ramachandra Guha 1989b). Often the women joked and said that "stealing from the forest is a tradition *(riwaz)* here." By describing their reliance on the reserves as an activity driven by necessity that they have long practiced, women assert an entitlement informally and overlook the formal boundaries set by the Forest Department. Based on the memory of the practice of forest use and the pressure of subsistence, local villagers hence make a claim on forest products and draw their own lines of property and redefine those established by the Forest Department.

The narratives of access and withdrawal captured from memory or from subsistence are powerful idioms to persuade the forest guard and sometimes even higher Forest Department officials. The forest guard, who is expected to stop women and fine them for stealing, sometimes gives in to the moral discourse of survival espoused by the women. For instance, he once acknowledged that women are forced to go to the forest for their everyday subsistence and said,

> I know it is very hard for the women. I understand their hardship. My mother in my village does the same. I know they have to do this. I know they cannot pay but I have to fine them once in a while, otherwise whatever little *fear of patraul* exists will also disappear. (16 January 1993, field notes; italics mine)

Even the divisional forest officer of the district reluctantly acknowledged that women are desperate and risk their lives to go to the forest. Although the recognition of women's demands for subsistence does not translate into formalization of rules of access to the forest, it creates possibilities to maneuver and modify rules of the Forest Department. Women recognize and manipulate the department's ambivalence and try to convince the forest guard either by logic, deceit, or even force. The practices of forest use are evidently embedded in a culture of conflict, compliance, and complicity in which the internal contradictions of property are enhanced. Property hence emerges as a complexly enacted and socially constituted practice of power. To further elaborate the contradictions of property, in the following section, I discuss the politics of community with the aim of disaggregating community as a homogeneous unit and describe the ways in which property is practiced along the lines of power and conflict.

Like property, the concept of community circulates as an orthodoxy in the circles of common property. The notion of community is repeatedly employed as a site where a culture of communitarian values, shared knowledges, and shared resources prevails. The strategic use of the concept of community may be useful to assert identity claims over resources, but more often than not, the concept of community coalesces and obscures the relations of hierarchy, inequality, and exploitation that circumscribe it.[8] Like the popular concepts of local and indigenous, community too requires a thorough interrogation that dismantles it and uncovers the layers of power that constitute it. Hence an ethnographic look at community through the lens of local history and politics, kin networks, and landscape allows us to move beyond a simplistic treatment of community and helps us describe how power operates in day-to-day life and reconfigures community and subsequently property.

In Bankhali, women are primarily responsible for collecting fuelwood and fodder. They spend many hours in the forest to meet the demands of subsistence. They climb tall trees and risk their lives for the sake of fuel and fodder. Tales of women losing their lives after a fall from a tree and of broken backs and legs abound in Bankhali, yet women go to the forest regularly. Because the competence and strength of a good mother and wife, especially newly married ones, are judged by the amount of wood they can carry, younger women are forced to make frequent trips to the forest and carry ever more wood and leaves. They compete with each other, constantly comparing the height of their fuel stacks and chastising those whose stacks are low. Women's responsibility to support the hearth is rooted in patriarchal relations of control and authority and forces them to bear the burden disproportionately. Women work harder and longer, and their husbands, brothers, and sons only occasionally partake the responsibility the women bear. That women are forced to go to the forest obeys the principle and structure of patriarchy and control. Women make subsistence possible for their entire families but are unable to inform the way in which decisions regarding forest management are made. The relations of unequal burden of work, patriarchal domination, and lack of control over land or hearth implicate women in hierarchical and exploitative relations of power, countering the idealized notion of community.

The gendered relations of property and community emerge more clearly in the panchayat forest (Agrawal 1994c; Sarin 1995a). The panchayat forest are managed by a council of villagers, but women who cut leaves and collect fodder have no say in decision making. On the contrary, men believe that women are uneducated and unaware of the importance of forests and hence do not understand the complex mechanisms of regeneration and protection. Ironically, even when men acknowledge that women are forced to go to the forest to meet the demands of their families and children, they hold them solely responsible for destroying the forest. For instance, when the Bankhali ban panchayat opens to the villagers for cutting branches and pruning *(safan)* for a week or two every year, the decision to open the forest and the time to do so are determined by the men alone. In fact, more than once, women disagreed with the decision of the council, but the decision of the male council prevailed. Also, for instance, when the panchayat forest opens, women climb tall trees, but the men, mostly the *panch* members, walk around to check if the women are cutting the trees appropriately. The men are convinced that women, even though they spend long hours in the forest, are incapable of understanding the principles of conservation. Men loudly instruct the women on how to cut leaves and not to harm the main tree trunk.

According to the rules of the Forest Department, there should be at least one female panch in the five-member panchayat body. In all three ban panchayat committees in Bankhali, there is a female panch member; however, her presence and participation exist only on paper. None of the female panch members ever actively participated at any level of decision making or attended any meeting.[9] While the female panch is never consulted by other members of the panchayat, other male nonmembers take a keen interest in the affairs of the panchayat and actively participate in the proceedings of the meeting. Women say that they cannot talk to the forest guard or to the members of the panchayat, as the patriarchal relations of kin hierarchy and age restrict women from speaking to men, and especially older ones. Evidently, as Bina Agarwal points out, "property under State, community, or clan ownership remains effectively under the managerial control of selected men through their dominance in both traditional and modern institutions: caste or clan councils, village elected bodies, State bureaucracies at all levels" (1994, 13). The structures and relations of patriarchy, hierarchy, and conflict suggest that the relations of property are determined not solely by who owns property but more importantly by traditional and institutional arrangements that regulate who *controls* property. Because men in patriarchy traditionally govern and dominate all structures of power, the rules and rights of property are forced to succumb to the contours of that power (Agarwal 1994; Rangan 1997).

Like gender, caste and class appear as important axes along which resource use patterns are shaped. The narratives of forest use in Uttarakhand present a homogenized picture of forest users, glossing over the differences and hierarchies that exist between them. There is a tendency to view rural India, especially the Himalaya, as a model of equality and unity. The near absence of caste and class hierarchies in Uttarakhand, and more specifically in the Kumaon Himalaya, is highlighted (Sanwal 1976; Ramachandra Guha 1989b). Guha writes:

> The (attenuated) presence of caste notwithstanding, hill society exhibits an absence of sharp class divisions. Viewed along with the presence of strong communal traditions, this makes Uttarakhand a fascinating exception which one is unable to fit into existing conceptualizations of social hierarchy in India. (1989b, 14)

The largely egalitarian social organization of Uttarakhand helps construct an image of the Himalaya as an archetype of authentic India where social hierarchies were and are insignificant and provide the basis for sustainable resource use.[10] Contrary to Gadgil and Guha's claim that "diversification and territorial exclusion helped minimize inter- and intra-caste competition" (1992, 103), caste relations in Bankhali and neighboring areas present a different picture. Upper-caste positions allow greater participation in resource decision units and encourage alliances along caste lines. In the six panchayat forests I studied, the *sarpanch* (head of the council) was from the higher caste, but the difficult task of guarding the forest was given to the lower castes. The lower castes, even when they hold positions on the panchayat committee, tend not to speak up or actively participate in the decision-making process.

The women's narratives of resource use are also caste and class based. The Brahmin women claimed that the Thakur women took a greater share in the forest because the social controls among the Thakur caste are less strict, allowing them to go into the forest after sunset and boldly confront the forest guard. Conversely, the Thakur women made fun of the Brahmin women, saying that they spent their entire days in the forest and were never satisfied with what they had. One of the Thakur women discussing the state of the forest said, "They [Brahmin] want more and more. I don't know how much! Brahmin stomach never fills up. [It seems] they want the entire forest." The Brahmin and Thakur women blamed each other, but there was a kind of consensus among both the Brahmin and Thakur villagers that the real culprit in the destruction of the forests was the Shilpkars, the Scheduled Castes. The Shilpkars were unequivocally blamed, as they were said to cut wood not only for their daily use but for sale in the town. Because the Brahmins and Thakurs collected fuelwood and fodder for subsistence, augmentation of income from forest products was seen as unscrupulous. The Shilpkars, in turn, repudiated any such accusations and pointed out that the forest guard was a Brahmin and favored the Brahmins while treating the Shilpkars harshly.

In addition to the dynamic of caste, class, age, marital status, position in the household, and kin status are all important factors that influence the politics of forest use. Women whose husbands or sons were employed as schoolteachers, drivers, or government employees, and women who were educated and were schoolteachers or worked in local day care centers and earned a salary, did not depend on the forest as much as those who had little or no access to income from outside. Moreover, the villagers employed in the state department were able to better understand the rules and regulations and occasionally appeal their cases in the head office. For instance, the villagers with jobs were the first to acquire cooking gas connections not only because they registered before others did but also because they were able to use their connections to jump the long queue. A statement of an older untouchable man summarizes the culture of collaboration and competition that is based on the politics of class and caste and constantly mediates the claims of property and divides the community (see Robbins, and Jackson and Chattopadhyay, this volume):

> Nobody cares about a man who can neither read nor write. I am
> a thumb-printer (angootha chhap). If the patraul catches me, I can
> neither appeal nor oppose. I am blamed for who I am.

Evidently the community is fragmented along the lines of caste, class, gender, and sometimes age and marital status. The rules and codes of property are hence subject to negotiation and manipulation along these socially inscribed relations of power and describe the multiply mediated context of property and community.

Although the local politics of the community pose their own limits of control and conservation on the forests, the extent to which the rules of property can be enforced depends on the practices of monitoring. In the following section, to explore the complex ways in which the forests are and can be guarded, I focus on the practice, perceptions, and narratives of the forest guard.

FEAR AND AUTHORITY IN PROPERTY

The forest guard plays a critical role in the maintenance and enforcement of rules in the forest. To the villagers, he represents the labyrinthine world of bureaucracy, an image he works hard to maintain. Usually forest guards are formally educated and have completed secondary school education. After joining the Forest Department, they are briefly provided elementary knowledge about forest conservation, fire prevention, and maintenance of nurseries. Their primary job is to patrol their assigned forest range in the reserves, catch the offenders, communicate government rules, prevent and extinguish seasonal forest fires, and plant nurseries. Villagers interact with the guard most closely in his role as the forest watch guard and constantly debate his commitment to the job. There were many opinions about the competence and incompetence of the forest guard. Although some regarded him as alcoholic and negligent, others were more sympathetic. For instance, one woman said, "What can that poor thing do? After all he is doing his duty. He is answerable to his supervisors. He too has to show that he is doing his job, while we do ours!" However, according to the lower-caste men and women, the guard let the upper-caste women escape but never let lower-caste offenders go. Similarly, the men believed that he was not strict enough with the women, and the women insisted that the men bribed him.

The forest guard was fully aware of the activities of the villagers and pulled out his record for 1992, in which he had confiscated 930 sickles, but expressed his inability to single-handedly guard the entire forest twenty-four hours a day without any assistance or without any protection (shotgun). He said he knew that women carried two kinds of implements and gave him the rusted, unusable one that they did not bother to collect, hence his rising stock. According to him, women did not know the "right way" to collect fuelwood and fodder and hence were primarily responsible for the dismal condition of the reserved forest. He blamed the women because women not only cut down oak branches and leaves but in their haste also destroyed young trees and saplings. In his eyes, the village women were uneducated and lacked awareness of environmental conservation and did not care if the forests disappeared. Lower-caste villagers were to be blamed more because they were more desperate, less educated, and less informed about the crisis of deforestation.

In a discussion with five forest guards, current and retired, during my fieldwork in 1993, I found that the guards repeatedly mentioned how the relationship of the guards with the villagers has changed over the years. A retired forest guard in his eighties, recounting his experience of forty years, observed that the growing competition for fuel and fodder has made women fearless. As the women struggle hard to meet their subsistence needs, they do not fear the forest guard as they did in the past. One forest guard, pointing to his uniform, remarked that the uniform no longer meant anything to the villagers.

The discussion of fear, and its lack thereof, leads us to ask how forests are guarded, who has the authority, and what erodes it. The question of monitoring and enforcement of rules is tied to authority, and as Herring points out, "Authority is at the bottom of much of the debate about nature as well as the conflict in and around nature systems, even if buried in discussion of interests, stakeholders, and institutional design" (1998, 2). Authority, as in the foregoing case, is articulated through the language of fear, fear of uniform or of fines. The forest guards regret that women no longer fear them, and as one said, "The women are not scared of me; I am scared of them. I go unarmed but they come to the forest with all kinds of weapons. They come to fight with me. Where is the fear of the guard now? There [a neighboring forest] the women tied the forest guard to the tree and left. They are not scared of us!"

Authority is derived from the position of the forest guard as a state employee who has the power to threaten, censure, and fine women. But authority is also derived from the forest guard's access to, and command over, specific knowledge of the forest, which he gained during his training period (Herring 1998; Gold 1998). The forest guard repeatedly, for instance, focused on the low educational level of the villagers, especially women, and referred to them as uneducated *(anpar)*, ignorant *(na samajh)*, and backward *(ganwar)*. He regularly used English words and phrases to talk of forest conservation and the devastating consequences of deforestation. He described his number of years of experience and different parts of the Himalaya suggesting his breadth of knowledge and showed off his close ties with the officers. By describing his educated status, scientific rationality, and urban connections, the forest guard was able to assert his superiority over the local villagers and exercise his authority. But the authority of the forest guard is not uncontested.

Villagers questioned his competence and commitment and in fact ignored the authority he wished to communicate. Village women described him as young and ignorant, someone who does not have the years of experience, or an outsider who did not understand the demands of the region. Some believed that he took bribes, others said he did not do the rounds, and others considered him lazy. In their eyes, the forest guard was primarily a government employee who has no real stake in, and understanding of, the forest and its products. In a context where the discourse of corruption pervades every aspect of Indian social and political relations (see Gupta 1995), the forest guard's authority is challenged in the realm of everyday practice.

Although the women were able to confront and even threaten the forest guard on the grounds of his incompetence and ignorance, the panchayat forest presents a different scenario. Because the guard for the panchayat forest, referred to as *chowkidar* (not patraul), is a member of the village, elected by the villagers as a member of the council, villagers are more willing to accept his authority. Rarely did women attempt to dupe the panchayat forest guard. The fear of humiliation in the hands of a fellow villager and possible dismissal from the panchayat membership was a strong deterrent. The panchayati forest guard was attributed more authority, irrespective of his caste, as he was a local member who understood the villagers' demands and pressures, unlike the forest guard who was employed by the Forest Department and was merely "doing his duty." The fear of the current panchayat reflects the power that the long-practiced traditional panchayat system called *lath panchayat* (baton panchayat) commanded. The traditional recognition and respect for the panchayat in the past reproduces structures of prohibition and control that are socially acceptable and approved. The punishment and fines imposed by the forest guard bore a heavier financial cost, but the censure and fines by the chowkidar had deeper bearing. In that sense, authority as a social mechanism of monitoring is also culturally mediated. Even though derived from fear, the power of the two guards becomes meaningful only in the context of social relations and regional ecology.

Interestingly, the forest guard, despite the dictates of his duty, is responsive to the social and ecological context in which women disregarded his authority. As mentioned earlier, forest guards were sympathetic, and as one of them said, "If they [women] don't go to the forest, where else will they go to feed their children and families?"[11] The forest guard is a representative of the Forest Department and is expected to enforce its rules, but in day-to-day practice, one finds that enforcement and monitoring too are informed by the cultural politics of power that create possibilities for negotiation of formal rules of property established by the state.

CONCLUSION

I have focused on the day-to-day practices and discourse of forest use. By looking at the long-practiced exercise of procuring fuelwood and fodder from the reserves in Uttarakhand, my aim has been twofold. First, by focusing on the everyday practices of villagers, I want to push the concept of property beyond its impress as a prescriptive design. Property is coded in the language of rules, regulations, and boundaries aimed at regulating the behavior of the resource users, but property becomes meaningful, or not, only in the realm of everyday action. By emphasizing the practices that contest the dictates laid out by rule—that is, stealing and rule breaking—my attempt has been to rethink property beyond its legal and prescriptive confines and redefine it as a set of relations of power and practice. That people steal and break rules is well recognized and has provided ammunition for rule makers to reinforce control of the resource users via stricter fines and punishment. Rule violations have also been of interest to practitioners of common property who suggest mechanisms of monitoring, sanctions, and incentives to improve rule making. But generally rule breaking is viewed as being caused by a weakness in rule making, which can be corrected. By tracing the narratives that endorse rule breaking, I have shown that practices of property and rule breaking tell a story different from the one told by rules and boundaries, which are important to understand the struggle over resources. Importantly, rule violations describe competing and coexisting discourses and practices of claim on the forest. Based on the demands for livelihood and landscape, the history of access despite restrictions, and cultural politics of caste and gender, villagers assert their claim to these resources, although informally, and strategically overlook the rules that forbid them to harvest forest products. The everyday practices of property bring to the fore the complexity that informs the operation of property regimes and force us to rethink property as a socially constructed and historically prescribed relationship and practice. For a full understanding of the politics of resource use, it is important that we contextualize practices of forest use in a complex web of livelihood, local politics, and regional and landscape particularities.

Second, the focus on practices of property also demonstrates the internal fractures that shape the relations of power among the community of resource users. The community is not a homogeneously constructed aggregate of mutuality and consensus; it is informed by the cultural politics of interests and motivations, hierarchy, domination, and inequality. The fragmented domain of community relations suggests that special attention be paid to the way we conceptualize community-based management and take into account the many ramifications of power that inscribe relations of community.

The recognition of property as socially embedded and enacted is relevant in the policy arena as well. Because the boundaries and rules do not always translate into practice, the suggestions of decentralization, community-based management, or comanagement will not yield results unless these efforts are responsive to local histories and landscapes in which these practices are located (Sarin 1995a; Agrawal 1996a). Therefore, community-based management is a question not only of empowering different actors, formulating better rules, and introducing incentives but also of incorporating and understanding the claims and context of resource use.

Finally, the recognition of property as socially embedded and enacted is of relevance in the policy arena. Since the boundaries and rules do not always translate into practices, the suggestions of decentralization, community-based management, or co-management will not yield results unless these efforts are responsive to local histories and landscapes in which these practices are located (Sarin 1995a; Agrawal 1996a). Therefore, community-based management is not only a question of empowering different actors, formulating better rules, and introducing incentives but also of incorporating and understanding the claims and context of resource use.

NOTES

1 Uttarakhand is the collective term for the Central Himalayan belt of the Kumaon and Garhwal Himalaya, comprising eight districts: Almora, Nainital, and Pithoragarh in Kumaon, and Tehri, Pauri, Dehradun, Uttarkashi, and Chamoli in Garhwal.

2 See Somanathan 1991; Agrawal 1994c, 1996b; Sarin 1995a, 1993; Jodha 1986.

3 See Macpherson 1978; Bromley 1989; Feder and Feeney 1991; Ostrom 1990; Schlager and Ostrom 1992; Berkes 1989; Berry 1988; Li 1996; Rangan 1997; Ribot 1997; Vandergeest 1996.

4 The civil forests, the third category of forests, are under the control of the Revenue Department. The ongoing conflict between the Forest Department and the Revenue Department has left these forests in worse condition and even allowed cultivation and encroachment.

5 The name of the village has been changed to protect the identity of the villagers and the forest guards who openly shared their views with me.

6 Uttarakhand is faced with a high rate of male out-migration. Approximately two-thirds of the men (ages sixteen to fifty) migrate to cities as either skilled or unskilled workers. The extent of migration is reported to have increased over the last thirty years (Ramachandra Guha 1989b, 147).

7 The local Uttarakhand economy has largely been augmented by extensive trade and reliance on commercial extraction of forest products since the precolonial times. Several trading fairs that attract merchants from far and wide are still quite common (see Gururani 1996; Rangan 1993). Despite the commercial extraction of forest products, women's claims on the forest reserves highlight the need for subsistence over sale.

8 See Sinha, Gururani, and Greenberg 1997; Baviskar 1995; Rangan 1993; and Agrawal 1995 for a critique of the use of community in the Indian environmental discourse.

9 Hobley (1991) describes a similar situation in a Nepalese village where the women did not participate in any decision related to the forest. A handful of men decided on behalf of the entire village.

10 See Sinha, Gururani, and Greenberg 1997; Gururani 1996.

11 The explanations and interpretations of deforestation and degradation offered by the forest guard did not differ much from those of the village women. He described deforestation as a cyclical phenomenon—forests appear and disappear, and humans cannot do very much (*parikasht,* predetermined). Also see Springer's (this volume) discussion of the "blurred boundaries" of state and society, where the agriculture officers, like the forest guard, rely on both local and scientific knowledges, contesting the neat separation that describes them.

Paul Robbins

Pastoralism and Community in Rajasthan: Interrogating Categories of Arid Lands Development

Participatory and community development are the hallmark tropes of an emergent "people-centered" sustainable development (Rahnema 1993; Esteva 1993). In arid India,[1] where many previous development efforts have floundered and failed, the language of this approach takes distinct forms. Here, images of community are strong; desert village communities are seen as coherent social spaces of reciprocity and contract. The image of the pastoralist is equally strong; members of identifiable, semi-itinerant, specialist castes are envisioned roaming the margins of the agricultural economy. These marginal groups, themselves seen as a coherent community, are understood to be undergoing a transition to commodity exchange relations, their social bonds dissolving. Taken together, this image of the development process is monolithic: people called *pastoralists*, living in social settings called a *community*, are undergoing economic transformation to something called *capitalism*.

The actual complexity of agro-pastoralism in India demonstrates the fractured character of local political economy and defies such conveniences. Following the work of other contributors to this volume, I will argue here against the reductions employed in these characterizations of village reality and maintain that these categories and the socioecological models they fit are ill suited to the situation of producers in arid lands. The tensions between local communities and the state, shown in the chapters of Saberwal and Rangan (this volume), often rival and articulate with local struggles themselves, as shown in the chapters of Guha and Gururani (this volume). Applying this to the status of agro-pastoralists in Rajasthan, I will show that communities and pastoral practices are diverse and divided, thereby denying the assumption that village localities are the inherently cohesive or natural scale at which to focus progressive development.

This chapter is divided into four parts. I begin with a review of the meaning and use of "community" in afforestation and wastelands development in India's arid zone. I then examine the deployment of the term "pastoralist" as it is used to construct producers in the region. In the next section, drawing on secondary sources as well as primary research data from villages in western Rajasthan, I assess who really constitutes the pastoral communities of arid India.[2] Finally, I examine the upheaval of adaptive production patterns in recent years and explore how community is constantly dissolved and reformed in response to market and land use change. In summary, I argue that community, like class and gender, is a *process* that, when understood, allows interpretation of economic change and facilitates more equitable agro-ecological development.

DEFINING COMMUNITY

The Indian development paradigm, built around the simultaneous urge to foster decentralized village democracy and to build coherent central planning, has constructed several colliding models of community. These models form policy and become inevitably intertwined with living village politics and economics. These discourses thereby become the process "through which social reality inevitably comes into being" (Escobar 1996, 46). Drawing on well-entrenched notions of village life, contemporary arid lands planning discourse offers the familiar and easy-to-accept images of communities as (1) organic societies, (2) moral economies, and (3) social contracts.

Organic Societies, Moral Economies, and Social Contracts
A burgeoning interest among ecological development offices in the Indian government surrounds "community-based" development and "social forestry." These programs represent an important response to paternalistic approaches in natural resource development and management. In the arid zone, these programs have been slower to emerge than in other regions but

are making recent headway, especially in the form of "Community Waste-lands Development" (Dhiman 1988). "Social Forestry" programs, especially those geared toward "the weaker section" of the economy (A. L. Rao 1988), have also recently begun to follow these forms of development into the arid zone.

In a drive to shift from expert power to local power, these programs have posited a local village community to whom a central authority needs to communicate and respond. To achieve people's participation in Rajasthan, foresters seek to establish, for example, "TRUCO. . . a combination of trust and confidence. . . . among the village *community*" (Dhiman 1988, italics mine). This trust is to facilitate development efforts, specifically, the plantation of fuelwood trees and the conservation of soil for the good of the community. In general, this approach treats the community or villagers as a natural and unproblematic category. That disempowered communities exist within the village is largely unacknowledged. Similarly, that a range of differing production strategies, from agriculture to pastoralism, stands to lose or benefit differentially from species plantation and enclosure decisions is also left unsaid. Instead, wasteland development strategies approach the village community as a whole.[3] In contemporary planning, the "village republic" continues to reign, at least on paper.[4] This orientation toward a single and organic community is mirrored in a sociology of village life founded in a moral economy.

The redistributive structures of obligation and debt, often referred to as the "moral economy," provide an underlying model of village life for community development, especially in the arid regions of India. This model of village economy emphasizes that exchange, storage, and investment are all situated in community-wide institutional systems that spread risk and "share poverty" (Scott 1976; Halperin 1994). In India, systems of semi-feudal exchange (Wiser and Wiser 1963) and community property rights (Bandyopadhyay 1983) are offered as evidence of systems of reciprocity embedded in social structure, caste, and community identity. For arid region development planners and observers, this model of the community is evidenced as the apparent integration of diverse caste interests through reciprocal obligation. In development analysis of village Rajasthan, for example, "interdependence, the division of labor, and the principle of reciprocity in economic exchange" explain "the village hierarchical system as a whole" (Sharma 1992). In a similar vein, the ecological linkages between varying kinds of production (agricultural and pastoral, in particular) are often posited in functionalist terms (Srivastava 1991).

For planners, this conception becomes a good foil for criticism or reform. In critical approaches, the moral economy is a lost or disappearing past that must be either shored or replaced (Chambers et al. 1989). Alternately, economists advocating increased security for private property rights see the moral economy as either a barrier to or a facilitator of growth (Hayami and Kikuchi 1981). In either case, whether help or hindrance, the community is viewed as a moral economy, a set of exchanges, and a stable system.

In another incarnation, the village becomes the location of individuated producers acting in enlightened self-interest and achieving community through the calculation of individual benefit. Defined in the tradition of institutional economics, such communities are the establishment of "collective action in control of individual action" (Commons 1990, 69). As the emerging dominant paradigm, this rational actor approach has come to dominate development discourse. This perspective is most prevalent in the common property approach of Bromley (1992), McKay and Acheson (1987), Ostrom (1990, 1992), and others. For common property theorists, the key to creating cooperative action is to understand individual producer benefits from collective resources and to create systems of monitoring and enforcement that protect them and the "social capital" that holds them together (Ostrom 1992a, following Coleman 1988).

In arid lands development literature, this approach has made large inroads. Plans for the revival of the local governance are phrased in terms of commons, tragedies, and incentives (Saxena 1994). Participatory development is increasingly dominated by rule-crafting language as an alternative to market-based and central-planning-based discourse (Chopra et al. 1990). Since the revelations of Jodha (1985, 1986) and Brara (1987) during the 1980s about the importance of common property resources to the rural poor of Rajasthan, this paradigm has become especially prevalent. Even where arid land developers are more critical of the village community, they continue to invoke the language of contract. Weighing the value of social forestry, one regional planner invoked the image of "millions of people collecting headloads from forests," warning that social forestry was inadequate and that "the restoration of discipline, therefore, is a must and the sooner it is done, the better for our survival" (B. Agarwal 1988, 177). The village community, as a whole, is here in breach of contract and must be "disciplined." No matter that many groups use and exercise rights while others may not, nor that many producers, as we will see, extract illegally in an act of resistance against hegemonic authority. The community is still a monolithic whole, unified in contract. Like the organic community and the moral economy, the socially contracted community is undivided and makes the logical target for development.

Why does this story of the unified community persist? In considering this question, it is important to bear in mind that these various accounts of community are not necessarily mutually supporting. Indeed, the notion of an organic and natural community contradicts the atomized structure of the social contract. The three notions, grounded in separate discursive histories, come to govern the planner's view of the village not because they are a logically coherent argument, however, but rather because they form a compelling narrative. In their diversity, they satisfy the contradictory demands placed on the image of the village. The community may be organic or economically logical by turns, as necessary in the development process. The moral economy may be less an actual characteristic of the past than a "meaningful image of village history, growing from the struggles of economic change" (Roseberry 1989, 59). In the same way that "development" is reinvented and reconceived while remaining steadfastly the same (Esteva 1993), so too is "community" made powerful by its ambiguity. It remains difficult to argue against development strategies that are "community based." Who, after all, does not want community?

Communities of Difference and Struggle

The rise of this triumvirate of constructions of the Indian community has not gone uncontested. In critical development literature, feminist scholars and other researchers in political economy challenge this naturalized view of community. Young (1990) questions the normative ideal of community as a useful starting point for political action, emphasizing that the very urge to establish and define community serves to marginalize minority and disempowered groups. Li (1996) convincingly argues that "community" is a strategically deployed category, rather than an empirical reality that can be used to politically unify or divide groups and locations. In arid India, the unified community is increasingly called into question. Ongoing division of community interests is still sometimes invoked as a fundamental force in environmental change (Goldman 1991), and differential access to power and capital results in differential management interests in irrigation and agriculture (Leaf 1992). Similarly, the choice of community-level development in south Asia often results in the exclusion of women and the loss of extant rights (Chen 1993; Sarin and Sharma 1993; B. Agarwal 1988). Even where women are incorporated into the local development approach, women from marginal castes or tribes tend to be ignored and passed over (*Economic and Political Weekly* 1992). These approaches dethrone the primacy of "community," replacing an explanatory economics of totality with an exploratory geography of difference.

In pursuit of this goal, Leach (1991) and Li (1996) define and identify the community as a micropolitical economy where division and unity move across a constantly changing political map of social process. Similarity and difference are negotiated, hijacked, and reformed along the fault lines and schisms of local power (Morrison and Bass 1992, 92). An examination of this kind of micropolitics in arid lands development, therefore, offers a response to the hegemony of the persistent story of village social life: community. In the same way, the approach must address the dominant trope of arid lands production embedded in the normative concept of "pastoralism."

DEPLOYING PASTORALISM

Like community, "pastoralism" is a term made problematic by its apparent transparency. The material practice of animal raising is usually understood to give rise to movement, social structure, and political affiliation in an obvious or unproblematic way. In much of the literature, pastoralists are seen to have "interests," and "ways of seeing" the world. Yet where contemporary pastoral practice is encountered in reality, it brings into doubt many commonly held notions. Without rehearsing the vast and sophisticated literature that seeks to define and frame the category of pastoralism,[5] it is sufficient to note that definitions are subtle, complex, and contested. Even so, pastoralism remains a largely unscrutinized category for state planning and development.

Colonial and Planning Accounts

Colonial accounts of pastoralism in northern India established clear categories for what and who constitutes a pastoralist, despite the incredible diversity of actual pastoral practice under colonialism. The underpinning of the pastoral category was generally caste based, as seen in gazetteers (Rajputana Gazetteers 1908) and censuses (M. Singh 1894) of the era. The urge to define pastoralists, linked with the urge to discipline and control them, flew in the face of the heterogeneity of agro-pastoral activities in that period, however (Bhattacharya 1995). Dependence on pastoral resources actually grew and diversified during the colonial period while animal keeping became more generalized under colonial rule. Even then, the dynamic quality of the pastoral sector entirely eluded colonial definition where a continuum of strategies prevailed and in which many groups, not just caste specialists, were involved (M. Jain 1994).

Present-day accounts follow a similar pattern. Although they do not always confound pastoral with "nomadic," in using the term "pastoralist," development planners continue to invoke and imagine an itinerant people primarily engaged in seasonal herd movement. In India, this is evident in an ongoing interest in traditional castes of pastoral specialists with predominantly nomadic adaptations (Sopher 1975; A. Prasad 1994). From this configuration grow the political imperatives of development planning: to either settle or facilitate movement. Development policy for pastoralists in the 1950s and 1960s favored gradual settlement under "Rehabilitation and Development Programs" in the second five-year plan (Mahapatra 1975). Recent policy has been more ambivalent toward nomadism and has, on occasion, attempted to facilitate it through interstate agreements and pressure (Kalla 1993). In either case, animal raising is conflated with movement. The implicit assumption that the interests of settled animal raisers will be met and subsumed under other sectors of development has proven spurious. Despite the massive growth in dependence on livestock, local-level development remains predominantly geared toward crop production (Henderson 1993). In the mind of the planner, villages are made up of peasants, and pastoralists live wandering on the margin.

The implementation of these neat categories, itinerant livestock raising on the one hand and intensive agricultural production on the other, reflects the simultaneous construction of "pastoralists" and "peasants" out of diverse agro-pastoral communities. Agro-pastoral practice, the mixed and adaptive emphasis of most desert producers, becomes a nonentity. This supports the effort throughout the peripheral arid region to secure state hegemony and facilitate capital accumulation through development. To plan a system of production and accumulation, the identity of producers must be fixed and known (Bhattacharya 1995). Further, agro-pastoral production practices and systems that represent precapitalist institutional formations and defy easy integration into capitalism (animal lending, for example) must be displaced. Categorical erasure of these ambiguities enables institutional and economic transformation. Seen this way, the categories of "pastoralist" and "peasant" are discursive deployments through which "state hegemony," as Henderson (1993) observes, "legitimizes the enterprise of rule and rationalizes capitalist production in the desert."

Toward Alternative Notions of Pastoralism
Globally, this monolithic view of pastoralism is being called into question. Seen as a range of material practices and ecological logics rather than as a state of mobility or identity, animal handling in India begins to resemble ranching in North America in its ambiguous relationship to capitalism, for example (Koster and Chang 1994), or to the state (Gilles and Gefu 1990). Ultimately, the practice of pastoralism is the exploitation of marginal, armed, and toxic plant species through the breeding and herding of livestock. Characteristics of the practice include a vast number of strategies including fallowing of land for fodder, animal lending, migration, and the exploitation of markets for meat, milk, and wool. Development problems experienced in such systems include use and access rights in community forest and pasturage, fodder availability and supplements during dry seasons and drought years, and access to livestock commodity markets and transport. Many producers experience these problems, not just traditional specialists.

This is not to argue that anyone who holds animals is a pastoralist. Emic views and categories of pastoralism do underline significant delimitations of pastoral identity (Westphal-Hellbusch 1975). Rather, it is to emphasize that the range of animal holding strategies is wide, diverse, and continuously overlooked in the creation of development policy. The proper or exact definition of *pastoralism* is less relevant here than the use of the term. Pastoralism is a notion deployed to rationalize, plan, and control

Table 1: Average Livestock Holding by Group

Caste	Cattle	Goats	Sheep
Brahmin	0	3	0
Jat	3.5	0	0
Meghwal	6	13.7	27
Nat	0	0.5	0.5
Raika	1.8	20.8	85.8
Rajput	4.1	9.1	19.9
Sindhi	4.4	9.6	53
Average Holding per Household	2.8	8.1	26.5

and is constantly defied by producers' adaptive capacities. Like the notion of community, pastoralism is defined through development discourse in a way that disciplines local economy and limits understanding of local complexity. An alternative approach to Indian pastoralism lies in the opposite direction. The charted terrain of pastoralism and community, as modeled by state planners, might be challenged by mapping economic and adaptive variability in agro-pastoral communities. The following analysis of agro-pastoral change in Rajasthan reflects such a geographic project.

WHO ARE RAJASTHAN'S AGRO-PASTORALISTS?

Despite images of continuity in depictions of pastoral community, diversity reigns in livestock production in Rajasthan. The animal economy spans a range of strategies from traditional specialists managing large herds, through nontraditional community elites with recent entry into livestock raising, to marginal households who depend heavily on a handful of animals for access to cash and protein. Each of these forms of agro-pastoralism is practiced throughout Gujarat and Rajasthan. The range of communities practicing some form of livestock management is shown in table 1, where livestock holdings are shown from a sample of forty-eight households in four villages in western Rajasthan. These practices vary considerably between caste groups and reflect different aspects of pastoral production.

Although caste is often an inappropriate starting place for the analysis of rural community interactions (Dirks 1992; Srinivas 1994), caste groups do continue to occupy distinct positions relative to capital, land, and state

Pastoralism and Community in Rajasthan 199

power. Traditional herding communities in Rajasthan—*raika* and Sindhi Muslims—continue to manage the largest herds. But to say that caste tradition influences current practice is not to say that it is simply determined. The Sindhi, for example, are cattle herders by tradition but now keep large numbers of small stock. Other nontraditional configurations are also in evidence. Sheep herding is increasingly an option for village elites (Jats and Rajputs) who traditionally shunned the practice. Even so, not all nontraditional communities have turned to animal keeping. Brahman households hold little or no small stock in keeping with the traditional sense that sheep and goat herding is an inauspicious practice. Marginal and land-poor groups (*bhil* tribals, and *meghwal* leather workers predominantly) have not made the difficult switch into large herd keeping but do keep a crucial handful of animals that are an important capital and labor investment. Several of these marginal households reported that they have switched into goat keeping for the first time this generation, having previously kept cattle to meet daily milk needs. The overall picture, that animals play a significant role in a variety of very different households, is an important one, but the variations in pastoral practice grow from the diverse characteristics of each of these groups. Specialist herders, elite investors, and marginal adapters each practice a different strand of pastoralism. I will introduce each in turn.

Pastoral Specialists

When distributing aid for pastoralists in the region, in the form of veterinary medicine or other materials, state and nonstate organizations usually target traditional pastoral specialists, especially the raika, Sindhi, and *gujar* caste groups. These specialist communities receive considerable academic attention and popular press.[6] Popular accounts notwithstanding, this recent research into specialist communities has revealed several important realities of pastoral life. First, it is evident that nomadism, while increasingly practiced, was not the traditional adaptation for pastoralists in the region. In Gujarat, migratory practice has been traced by Cincotta and Pangare (1994) to a set of evolving historical constraints on production where the changing availability of fodder, crop residues, manure markets, and labor led to constrained options in dairying, settlement, and migration. In Rajasthan, the incidence of migration also appears relatively recent but increasing (Köhler-Rollefson 1994; A. Prasad 1994). For Rajasthani specialists, the constraints of local fodder availability are coupled with the growth of wool and meat markets leading to increased annual migratory movements, particularly for sheep and goat herders (Robbins 1998a). In either case, migration is not undertaken lightly. The decision to migrate is generally made with reluctance and only under optimal conditions. The dangers, costs, and problems of migration remain daunting (Salzman 1986; Köhler-Rollefson 1992b, 1993). This reality belies the notion of Indian pastoralists as "nomads," geared historically around movement. These conclusions also depart from a broader global model that predicts the decline of mobility in arid land adaptations (Bose 1975; Salzman 1980).

Second, it is evident that the political structures of specialist pastoral practice are complex and held together by internal tensions and constraints as well as through the control of information (Agrawal 1994a, 1999). Pastoralists are increasingly seen as careful independent producers, linked through institutional systems of cooperation and monitoring into production groups for mutual aid. These groups represent producers with sometimes diverging interests joined together for collective benefit. The resulting economies of scale and divisions of duties and decision making benefit the community despite highly individuated producer interests (Agrawal 1993, 1994b). These investigations of motivation, strategy, and negotiation show the specialist pastoralist to be less like an idealized village cooperator and more like his industrialized counterpart, the rancher (Gilles and Gefu 1990). Additionally, it is clear that not all specialists are equally well endowed. It is traditionally the case in most pastoral societies that producers occupy a relatively slender range of capital holdings; in theory, pastoralists who become too wealthy or poor drop out of livestock management (Barth 1961, 1964). In Rajasthan this does not appear to be the case. Although a portion of the specialist community holds herds of unprecedented size, an increasing number of pastoral households have only a handful of animals and practice herding under marginal conditions.

Finally, investigation into specialist herding reveals an overwhelming market orientation. Camel-herding specialists are linked through seasonal markets to the trade in draft animals (Köhler-Rollefson 1992a, 1992b). Sheep herding is increasingly capitalized, turning the best profit by timing the sale of animals on migration near urban markets (A. Joshi 1987; Agrawal 1992, 1993). The growth in the size of the regional herd in recent decades is clearly linked to the growth in wool markets. Meat markets, while unacknowledged by development professionals and government officials, are also growing. The demands of urban meat consumption have also boosted the holdings of sheep and goats in specialist herds (Robbins 1998b). This market orientation departs from many images of pastoralists as subsistence producers (Köhler-Rollefson 1994).

In sum, the specialist herders of the region are increasingly nomadic producers, tied to a commodity-driven livestock market, with highly individuated interests coaxed into cooperative production for mutual benefit. In addition, the stratification of the specialist population from wealthy to very poor is an increasingly prominent fact. Significantly, these realities of pastoral production contradict several commonly held images of "pastoralism." Movement averse, class divided, and market oriented, the specialist herders of arid India do not fit the mold cast by development planners and many researchers.

The Entrance of Nontraditional Elites

Receiving less attention in the literature, the entry of nontraditional pastoralists into herding represents an entirely different manifestation of pastoralism. Little literature exists on this phenomenon, but a general sketch is possible. Predominantly coming from the elite Rajput caste who historically ruled in the region, and from the powerful Jat community that is usually associated with farming, these households have different opportunities than specialist herders. The unique capital, land, and political resources available to them deeply influence elite pastoral adaptation. Traditionally, these groups were in no way connected to animal production and broadly shunned contact with livestock, especially small stock (M. Singh 1894). The decline of caste-based professional specialization in the post-independence period has altered this arrangement significantly. As examined by Kavoori (1990), Rajasthani Rajput communities in the northern and eastern parts of the state are taking increasing advantage of the booming wool and meat markets by organizing large herds and migrations. In the west, it is not unusual to find herders from elite castes with herds in excess of three hundred animals. Their relatively large land and capital endowments allow them greater flexibility in migration timing and spacing. Interviews with animal-holding Rajputs suggest that elite producers are free either to conduct migrations themselves or to loan their animals to specialist herders for the duration of the migration season. If they should migrate, their political clout, not enjoyed by other migrating herders, is likely useful in arranging grazing on private land or state forests (Köhler-Rollefson 1994). The increasing presence of elites in pastoralism undermines key tropes in the composite image of pastoralists as poor, politically marginal, traditionally disempowered groups. Elite herders bring their political and material resources to bear on the problem of herding, and these groups are keeping increasingly large and productive herds.

Marginal Herding on the Periphery
Even less acknowledged than the position of elites in pastoralism is that of the most marginal communities. The poorest households in the arid region, representing a range of Scheduled Castes and Tribes, are dependent on livestock production, and the bulk of the booming goat population is held by the smallest and most marginal producers (Robbins 1994). The characteristics of pastoral practice on the margin reflect the limited resources and crushing constraints of this sector of the village economy. Lack of land resources, poor access to capital, and political disenfranchisement are each embodied in marginal pastoral practice.

Generally, poorer households increasingly rely on goats for the supply of milk and curd, which provides dietary protein, and for meat, which provides access to the cash market. Because of the traditionally low or negligible inputs of capital and labor into these small holdings, the efficiency of production for small stock under these conditions is remarkably high (Ahuja and Rathore 1987; Sagar and Ahuja 1993). Still, the management of small stock is not without serious risk and investment. Dry season fodder shortages are reported to be on the increase. Even marginal households often surrender the crucial labor of one or two family members for short migrations to nearby pasturage during the dry season. These most marginal communities are especially vulnerable to the water shortages, fodder deficits, and changes in the market that less affect other pastoral groups. Excluding them from the category of pastoralist, therefore, has practical implications. These households are rarely targeted for veterinary or marketing assistance. Nor are fodder bank projects or other pastoral development aid directed to villages dominated by meghwals, bhils, and other "nonpastoral" communities. The caste categories of the region disguise important economic changes and adaptations.

Women in Pastoralism

If "pastoralist" is a casted category, so too is it gendered. The image of the pastoralist in the region is overwhelmingly male as a result of the dominance of men in the public sphere of pastoral production. For Sindhi and raika households, women are not often present on migration. Land and animal tenure also seem to exclude women's control of key pastoral resources. Even so, women play a crucial and prominent role in pastoral production despite profound neglect by scholars and officials (Dahl 1987). Rangnekar (1994) emphasizes several often neglected roles of women in livestock production in northern Gujarat. The gendered burden of pastoral labor, the large share of gendered environmental and animal health knowledge, and the linkages between marriage systems and animal exchange all make women pastoralists important in production and politically more powerful than quick examination would otherwise reveal.

In Rajasthan, a similar pattern is evident. Although women are denied an explicit role in ownership, women often handle the marketing of many animals and animal products. Many families report that cash from animal sales goes back into household reproduction and not into the animal trade itself, suggesting the prominence of women in the animal economy. That women (and children) handle many animal maintenance tasks around the household, including the provision of forage and milking, also belies the image of the male herder. As long as the areas where women are less prominent, especially formal animal markets and migration, continue to be constructed as the sole locations of pastoral activity, women will remain invisible to researchers and planners. This follows the global pattern of analysis wherein women's tenurial rights, knowledges, and production roles are ignored in household and community-level research (Rocheleau et al. 1996).

In summary, the category of pastoralism, as currently deployed, elides the complexity of the human/environment interface and hides a range of socioeconomic practices occupied by groups practicing pastoral adaptations. Each of these groups holds a very different position in local politics and in the divisive struggle over resources that has become such a significant part of village life in recent years (Gadgil and Guha 1995). If we examine the diversity of adaptations to changes in environment and landscape, the category of "community" similarly falls into question.

Table 2: Changes in Land Cover, 1974–1994

Institutional Cover Type	Change (Average ha/village)
Waste	−142.2
Oran	−44.2
Fallow	80.5
Gochar	34.6
Enclosure	16.0

RECONFIGURING COMMUNITY IN THE ARID ZONE

The landscape of arid India, in which this broad range of producers is embedded, has undergone dramatic transformation in recent years. Intensification of agriculture, professionalization of forest management, and decline in traditional systems of land use have all served to transform the village environment. In adapting to these changes, pastoral producers have allied their interests in differing ways. Landholding elites and large herd holders have created new allegiances while less-powerful herders have joined marginal communities in protection of local resources. The result is an emerging political ecology marked by new alliances and divisions in the context of a resource base undergoing transformation and of a rapidly changing legal and biotic landscape.

The Context of Landscape Change
The institutional transformations that swept the region since independence have resulted in a dramatic reconfiguration of available resources. Table 2 shows the increase and decline in key land cover types from 1974 to 1994, based on a survey of village land records in twenty-eight villages in Barmer, Jodhpur, and Jaisalmer districts of Rajasthan. These reflect the more general trends in the arid zone since independence.

The waste category includes a large area of valuable grazing, browsing, and collecting land under an assortment of traditional management systems. The collapsing of these management regimes into the category of waste usually erases the record of extant management regimes (Brara 1987). *Oran* is village forest-pasture under the protection of traditional village sanction.[7] Gochars are grazing lands carved out of traditional management areas under the nominative control of the local *gram panchayat,* an elected village council. Fallow lands are long-fallow private holdings under grassy pasture coverage traditionally used as dry season grazing and open for all members of the village. Forest Department enclosures are public lands under plantation and closed to public use, especially grazing and browsing.

The two most significant patterns in cover change during the period are the trend toward state-centered institutional formations and the increase in fallow lands as a proportion of all grazing and browsing lands. The decline of "waste" land both represents and disguises the disappearance of a range of communal tenure systems. This trend follows the pattern elsewhere in the region and the country more generally (Chakravarty-Kaul 1996; Brara 1987; Jodha 1985, 1986). The decline of oran land marks the disappearance of traditional property systems of community use, monitoring, and enforcement. The decline of orans and wastes, each linked to high relative pasture productivity, also means a loss of crucial tree and grass coverage (Robbins 1998c, 1998d). The increase in Forest Department land and gram panchayat gochers marks the replacement of these traditional institutional configurations with central and local state tenures. Finally, the boom in fallow lands is a reflection of land reforms that continue to bring previously communal lands into individuated tenure. These lands are often unplowed or frequently fallowed but remain in private control nevertheless. These changes in institutional form provide the context within which producer adaptations occur and group alliances and negotiations are conducted.

The available resources give rise to varied production strategies throughout the year. Based on a cross-caste survey of land use strategies in twenty-eight arid-zone villages, these are summarized in table 3. An examination of variations in these strategies reveals some obvious trends. First, small holders and the land poor are highly dependent on community resources such as the gochar and oran. The general decline in these resources is devastating for marginal producers (J. Arnold 1990; J. Arnold and Stewart 1991). Pastoral specialists and elites are also more dependent on fodder feeding, migration, and especially the use of fallow lands. The strategy of marginal pastoral specialist households resembles most closely that of the land poor, especially in the dependence on local grazing land and the use of animal lending in the dry season.

These divergences and convergences in strategy lead to variations in sociopolitical village alliances, and struggles over village resources there-

Table 3: Herding and Collecting Strategies throughout the Year

Group	Monsoon (July–October)	Winter (November–February)	Dry Season (March–June)
Large landholders >20 ha	Fodder feeding	Own fallow, other's fallow, fodder feeding	Fodder feeding, hired migration,[1] long migration [1]
Small holders 2–20 ha	Gochar, fodder supplement	Own fallow, oran	Fodder feeding, oran
Land poor <2 ha	Gochar	Oran	Oran, animal lending, short migration [2]
Pastoral specialists	Gochar, short migration [2]	Other's fallow, oran	Short migration,[2] long migration [1]
Marginal specialists	Gochar	Oran, short migration [2]	Oran, animal lending, short migration,[2] long migration [1]

1. Represents a long and extended migration to adjacent states, especially Madhya Pradesh, Uttar Pradesh, and Haryana/Punjab.

2. Represents a one- or two-month migration to a nearby village or open pasturage within Rajasthan or Gujarat, usually directed toward regions where the producer has extended kin networks.

fore become arenas for coalition building and rivalry in ways that are classed, casted, and gendered.

Elite Alliances: Contractualization and Hegemony
The largest landholders in villages throughout the region come from elite groups (Rajputs and Jats) who have, to a degree, translated their caste power into class power (Omvedt 1978). This leverage of status into capital has been managed through manipulation of the land reform process, the management of informal loans, and the use of their status in "democratic" institutions such as the gram panchayat. The strategies of these groups have

an increasingly pastoral element and are linked to that of well-positioned pastoral specialists. Large landholders usually possess large herds, sometimes in excess of three hundred head of small stock (sheep and goats). These depend on fodder feeding during the growing season, on fallow grazing on their substantial holdings during the winter, and on fodder supplements, hired herding, and migration during the dry season as shown in table 3.

Elite pastoral specialists, while more dependent on community grazing resources such as the gochar, resemble their agricultural counterparts in many respects. The use of available fallow lands and the role of migration are similarly significant, and the closure of fallow fields during the rainy season sometimes necessitates short migrations for both groups. The capital required for both short and long migration is available in these households, as is labor. Agricultural and pastoral interests are merged through these commonalities of production in a way that reconfigures traditional caste relations. Despite important contrary case examples that emphasize the divergence of agricultural and pastoral interests (Agrawal 1994a), there is increasing convergence of interests between wealthy landholders and well-to-do pastoral producers.

The most apparent manifestation of this relationship comes in the form of contractualized grazing relations on fallow land. These contracts, while not written, are fairly formal and exclusive and displace the traditional obligatory systems prevailing in the early postindependence era. Traditionally, institutionalized systems of mutual obligation acted to keep fallow lands open for common village grazing during the winter and dry season. Closure of fields to the community traditionally brought punishments and reprimands from a group of village elders who represented a range of caste groups. The forum for the negotiation of these rights, the informal village council, was traditionally a strong and flexible authority with broad support across caste communities (Saxena 1994). It would be romantic to overstate the harmony of such an institutional structure, but it stands in marked distinction to the formalized and exclusive contracts of recent years. According to most herders and farmers interviewed in the region, an increase in exclusive contractualization began within the last twenty years. Individual households now contract to graze on the land of large holders for a specified period each year. In exchange, the dung from the herd is kept by the landowner. These households are also sometimes tied through animal lending arrangements in which landholders with large herds but little pastoral knowledge turn over their animals to specialists in exchange for fallow use and extra fodder. These pastoral households are sometimes historically linked to the landholder through preindependence patron-client ties. More often, however, these ties are new and regularly renegotiated. These arrangements ecologically mimic the traditional system of land use exchange where large numbers of animals deposit key productive nutrients in exchange for crucial winter chaff and perennial grasses (Srivastava 1991; Bose 1975). The social and political ecology of the contractual system is distinctly different, however, from the more open, traditional arrangement. Households with large herds are brought into a specified compact with large landholders, but marginal and small-holding herders are excluded.

The development of individuated land use contracts is paralleled by the hegemony of elites over local political mechanisms. Local statutory panchayats, centrally situated bureaucracies such as the Drought Prone Areas Program, and regional environmental authorities such as the Forest Department all act through political systems held tightly in the grasp of village elites. While panchayat elections have been highly contested in Rajasthan, the predominant struggle has been between Jats (elite farmers) and Rajputs (elite landlords) (Sharma 1992). In either case, the power of the largest landholders is paramount. The formation of Forest Department enclosures and other institutional developments are also influenced by local landholders.[8] Although the power of these local political institutions is somewhat limited, they do control the flow of important resources into the village environment.[9]

This elite hegemony has, in recent years, begun to reflect the interests of pastoral production. Collaborating with pastoral allies, farmers have gained increasing support for animal raising. In one case, the local Rajput landlord has gone to great lengths to bring veterinary support into the village. In another example, fodder banks have been proposed by Jat landholders. This shift in emphasis toward pastoral practice is not necessarily evidence of a new realization of pastoral needs for the desert environment on the part of state agencies. Rather, it reflects a growing alliance of interests between landholding elites and wealthy herders. Marginal producers are not excluded here, and they may inadvertently benefit from a fodder bank, depending on how it is implemented. Neither are they being planned for, however. A development orientation for pastoralism does not mean a development orientation for all pastoralists. These marginal animal raisers adapt through coalition building in the opposite direction, toward allied interests with other marginalized communities.

Marginal Alliances: Short Migrations and Herd Lending
A vast majority of marginal livestock holders in the region, no matter what their caste, hold fewer than fifteen goats and sheep.[10] As noted previously, these households are highly dependent on pastoral production, nevertheless. As access to orans and fallow land dwindles, new production coalitions emerge to facilitate short-term migrations and herd lending. In the first case, pastoral specialists with smaller or less productive herds may join the land poor in organizing short migrations. While group insularity remains something of a barrier to more widespread practice, and while these institutional formations remain more unusual than elite coalitions, they do occur with increasing frequency. More commonly, marginal pastoral households will assemble animals and resources from nonpastoralists, including meghwal and bhil households, to achieve the economies of scale that make even short migrations possible. In exchange, the animal owner may surrender the wool cut or some other animal commodity. This kind of animal lending is common and represents an emergent system of labor sharing. Although based on traditional management systems (animal loans are not new), these are significant because they introduce new exchanges into village economies and forge the basis for political coalition.

These kinds of coalitions on the margins of the village economy do not take the form of the political "iron triangles" created by elites, where bureaucracy, private interests, and the local state together monopolize development. Even so, the interaction of previously disparate groups does reflect emergent, albeit ephemeral, political identities. While traditional injunctions about interactions like the sharing of meals persist, and though direct organization remains invisible, members of meghwal, raika, bhil, and Sindhi communities do meet openly, exchange production information, and discuss village political matters. Group identity is reformed in slow and subtle fashion, re-creating the constellations of village power. The distribution and redistribution of power defy monolithic notions of pastoral community by creating patterns of difference and solidarity. In a similar way, the gendered character of ecological politics underlines the deceptive assumption of monolithic political identities.

Gender and Authority: Tree Cutting and Fodder Harvesting

Responses to the institution of gram panchayat and Forest Department authority over community lands highlights another side to community politics, that of gender. Specifically, women's protection of some community resources coincides with the destruction of others. Like ongoing problems with "women offenders" in joint forest management schemes (Sarin 1995a; Gururani, this volume), arid zone community pastures and forests commonly undergo heavy illicit cutting and resource extraction by women. Observation of Forest Department enclosures suggests that the extraction of fodder by women is the most common daily violation of local rules against tree cutting. In particular, women from households with animal holdings were significant "offenders." The significance of livestock dependence for many women is heightened by the absence of other significant property, and this provides a significant motivation for infraction (Chen 1993). Women usually spared village orans, however. In many cases, travel time across orans to other areas for wood and fodder harvesting was considerable. An essentialist portrayal of this effect might attribute women's choices to the "traditional" and "conservative" character of women's identity. This is not borne out by observation, however. In many cases, women cut orans heavily, even after long periods of careful protection, once panchayat or Forest Department authority was granted over them. In conversations with women livestock managers from several castes, many explain that they are largely dismissive of Forest Department regulations and complain vocally about the lack of panchayat accountability. Many women acknowledge that the upcoming panchayat elections (the first in thirteen years and the first with reserved seats for women) might change the ineffectual character of local state authority. Most remain skeptical, however, about the power and relevance of the institution. This discrepancy represents another significant "fault line" in village ecological politics. Women's extraction of state-controlled resources for pastoral production reflects the kinds of conflicts over representation and de jure versus de facto rights that mark many gendered environmental conflicts in other parts of India and throughout the world (Gururani, this volume; Rocheleau 1985; Rocheleau et al. 1996).

More to the point, like the reconfiguration of caste- and class-based production coalitions, gendered pastoral resource strategies are nested in a divided political landscape. Women's combined actions of extraction and destruction in gram panchayat and Forest Department lands represent Scott's (1985) "everyday forms of resistance" against imposed authority. At the same time, women's ongoing, combined protection of orans represents tacit coalition, across class and caste, to defend certain key agro-pastoral resources. Their actions serve to reinvent pastoral practice and to re-create community relations. Pastoralism and community are again shown to be fluid and fractured. What does that leave of these categories as analytical and development tools?

CONCLUSIONS: INTERPRETING COMMUNITY AS PROCESS

Despite their fragmented forms, pastoralism and community persist. They are formed through the action of producers in pastoral practice and through the struggle and exchange that marks community political economy. They are reflected in myriad actions: the seizure of market opportunities in meat and wool, the establishment of new exchange through contract, the opening of new production relations across caste lines. Communities are dissolved and reformed as pastoral practice is reinvented. Community is "drawn out of the mass" through positive reciprocity (following Hyde 1983, 38); where old bonds dissolve, they are reinvented in new forms. Pastoralism is re-created through ongoing adaptation; where old strategies become impossible, new ones are established. These alliances and material exchanges mark the political economics of contemporary agro-pastoralism.

The ambiguous nature of these processes is notable. Caste, for example, has not simply transformed itself into class. Instead, caste identity is fragmented and reformed under changing material and social conditions. Caste and class are interwoven and do not collapse well into each other. In the rise of new coalitions and the formation of strategies on the margin, unpredictable outcomes emerge, and village power has a "much more fluid character than is associated with either caste or class alone" (Béteille 1965, 187). Processes of exploitation and differentiation are apparent in the pastoral landscape of Rajasthan, without a single pattern emerging. Many producers do have less access to pasture and other key resources and are placed at a dramatic disadvantage relative to capital and to the instruments of village power. Economic transformation in agrarian India, as Gadgil and Guha (1992) observe, does result in the ascendancy of individualism, contractualization, and impersonal and rigid codes, along with the decline of traditional systems of resource management.

Yet despite this evidence of transition to production relations typical of capitalism, many noncapitalistic practices and institutions persist and thrive. The most vulnerable groups are engaged in adaptive responses to socioeconomic change. Coalition building and diversification of production strategies are apparent in even the most marginal households. Animal lending, fodder bargains, and joint migration all underline the range of socioeconomic practices and diverse institutional relationships. Thus pastoralism is making no simple or predictable transition to capitalism. Nor is it articulating into a new, single, or alternative form. Rather, it is unraveling into several parallel strands of production and exchange with noncapitalisms existing alongside and within capitalist relations. Like other complex economies around the world, community agro-pastoralism is "an economy that is not singular, centered, ordered, or self-constituting, and that therefore is not capitalism's exclusive domain" (Gibson-Graham 1996, 45).

The implication of this ambiguity for future planning requires much more thought. First, the notion of pastoral development clearly needs rethinking. Livestock orientation in development assistance does not necessarily serve the interests of the poor or marginal groups for whom such assistance is intended, though it might. Local research therefore becomes a crucial prerequisite to any program that stocks fodder, plants grasses, or gives away goats (Köhler-Rollefson and Rathore 1998).

Second, and more significantly, a truly progressive or emancipatory strategy must ultimately help to foster the emerging political alliances on the periphery. In so doing, it must avoid the reductive planning of "community-based" development. To that end, community itself might be redefined from a specific location or collective reality to an ongoing and negotiated *process.* By participating in that process rather than attempting to defy, control, or ignore it, development practice might move along more radical and practical lines. Rather than making the a priori assumption, for example, that a group or groups are marginal or empowered, that they are internally coherent or divided, or that they are pastoral or agricultural, NGOs (nongovernmental organizations) or other interested practitioners should investigate what divisions and adaptations are in place. Having identified coalition building and pastoral adaptation within and between marginal groups, practitioners might participate in adaptive pastoral change not only by providing direct assistance but also by facilitating and contributing to the process of institution building itself by asking new questions. "Who is trying to organize production with whom?" "What social and political barriers do such locally configured political and economic adaptations face?" Thus development practice would not "craft" idealized institutions or identify villagewide "social capital" as envisioned in some narratives (Ostrom 1992). Rather, it would acknowledge and participate in ongoing alliance building that might then, through ongoing negotiation of interests, eventually build toward all-village cooperative action. To arrive there, however, it is necessary to abandon the notion that such commonality exists before the fact. It also calls for a far larger research component for any such development activity. Pastoral development is political. The collapse of the neat categories of development such as "community" and "pastoralist" means that local intervention must be better based on research into the extant politics of adaptation and coalition, if for no other reason than to do less inadvertent harm than has often been the case. In this way, the death of traditional development narratives need not mean the end of emancipatory action. Indeed, it marks the beginning.

NOTES

1 The arid zone, as it is referred to in this chapter, refers to the semiarid and arid regions of Gujarat and Rajasthan, generally falling on the dry side of the 350 mm isohyet in the western portions of these states (A. S. Rao 1992).
2 The original fieldwork research for this chapter was conducted in twenty-eight villages in Marwar, western Rajasthan, during 1993 and 1994.
3 The image of an organic village is in no way unique to contemporary planners but is rooted in colonial (Inden 1990), nationalist (Mookerji 1919; Ghurye 1932), and modernist (Dube 1969, 204) planning and scholarship. This utopian vision of village community is also gaining currency with nongovernmental ecological activists (Rangan 1993).
4 This term takes a convoluted path to the present. Coined by Charles Metcalf (Inden 1990) during colonial rule, it was reappropriated by many nationalist scholars and later deployed in the neoinstitutional work of Wade (1988) and others.
5 See, in particular, Galaty and Johnson 1990; Dyson-Hudson and Dyson-Hudson 1982; Khazanov 1994; Galaty 1984; Johnson 1969; and Jacobs 1965 for more exhaustive discussion of typology, variation, and range of practices within pastoralism.
6 This attention is not always positive, but the earnestness of the development literature seeking to defend traditional pasturage and land rights (Salzman 1986) is mirrored in a distorted

way by the romance of accounts in *National Geographic* magazine. See, especially, Davison's (1993) account: "Wandering with India's Rabari."

7 These lands are protected through both religious sanction (Gold and Gujar 1989; Gadgil and Vartak 1975, 1981) and secular village monitoring and enforcement (Robbins 1998c).

8 See Agrawal 1994a for a related example where community lands were turned over to Forest Department authorities against the protestation of pastoral producers. The lands were closed by an elite-dominated panchayat with little interest in the grazing resource.

9 The limited role for these gram panchayats in determining development activities is well documented. But although technical support staff and funding are still poorly coordinated through these structures, they do continue to determine what little focus exists for local infrastructural development (L. Jain 1994).

10 Based on village-level figures compiled for the livestock census of 1992.

Labored Landscapes: Agro-ecological Change in Central Gujarat, India

"Patelón no varchasva dubva mandyo chhe" [Patel domination is beginning to sink]. I heard this refrain again and again while doing fieldwork in 1994 and 1995 in the subdistrict of Matar, an administrative enclave of the state of Gujarat in western India. I was there to examine how a massive surface irrigation project implemented in the mid-1960s had transformed the social and physical landscape. Had the advent of canal irrigation and a cash-intensive biochemical agriculture deepened existing economic inequalities? Was this attempted transformation of nature—its "appropriation" and "substitution," as some agro-food geographers describe it[1]—working to the benefit of the rich? The answers were a lot less clear-cut than I had expected. I found myself puzzling over that refrain I repeatedly heard: "Patelón no varchasva dubva mandyo chhe." The image that came to mind was that of a brilliant moon, its control of the night sky once undisputed, now on the wane. In fact, the Patels have been the dominant caste in the subdistrict of Matar and in most of Gujarat since the late 1830s.[2] They have controlled its economy and its politics. So why is their power now diminishing? Why didn't canal irrigation and the green revolution agriculture it made possible consolidate the Patels' hold over society?

These place-specific questions, which suggest deviations from a priori expectations, become significant within the context of a larger question: Why is there economic inequality within agrarian societies? Or for that matter, why is any society stratified? Although it may appear absurd on the face of it to attempt to answer such an enormous question, the fact remains that there is a general (by which I mean an acontextual) theory of inequality that is widely followed by scholars in the human sciences and is almost axiomatic within popular culture. In its simplest formulation, the theory maps initial endowments or asset ownership into future wealth. Although it does not propose a one-to-one correspondence between initial and future asset ownership—differences in luck, foresight, personal skills, targeted opportunities, and exposure to various sorts of contingencies introduce a stochastic element into the framework—the model does claim a high degree of positive correlation between initial and future wealth. In a nutshell, the model says that those who are initially wealthy have systematically more opportunities to become wealthier than those who are not, for a variety of reasons. Among the factors most commonly invoked (the first three are drawn directly from Kautsky's and Lenin's classic fin de siècle disquisitions on the "agrarian question"):[3] cheaper and readier access to credit, the ability to tap economies of scale in production and consumption, more bargaining power and hence the ability to negotiate superior terms of exchange in factor and output markets, better information about investment possibilities, denser social and political networks, and, finally, greater willingness and capacity to bear risk and capture the gains of innovation. The net effect? The cumulation of advantages over time, economic stratification, perpetuation of the status quo.

This basic model of inequality can be recast in the language of neoclassical economics, economic sociology, or variants of Marxism: in short, the three "grand" intellectual and political traditions that have molded the field of agrarian studies. And within the amoebic field of agrarian studies, the model takes its blunt form as the "linear proletarianization" or "polarization" thesis, which predicts that capitalism will eventually differentiate a peasantry into two classes—capitalist farmers and rural workers.[4] Applications of the thesis in this rudimentary form are increasingly rare. The more nuanced varieties of the thesis recognize the possibility of persistence of a smallholder peasantry or of family farms despite a general historical tendency toward polarization. Departures from the rudimentary model are couched either in terms of biological and ecological constraints within agriculture that inflate "turnover time," thereby impeding capitalist penetration,[5] or in terms of intrahousehold altruism that enables peasants to "super-exploit" family labor and outcompete capitalist firms,[6] or else in terms of a functionalist claim that capitalists permit smallholders to survive in order to maintain a reserve army of labor that keeps wages down and workers divided.[7]

The polarization thesis and its variants have supplied a powerful framework for understanding inequality in agrarian societies. It is precisely the framework that I carried with me to the field when I set out to discover the impact of canal irrigation and green revolution technologies in the subdistrict of Matar. I had expected to find a differentiating peasantry and growing inequities. Instead I found social flux, marked in some cases by striking reversals of fortune. The economic supremacy of the historically dominant Patel caste was under attack from groups that had until recently been the Patels' subordinates. Why?

The remainder of the chapter sketches an answer. In the process, I try to repudiate not only deductivist readings of history (such as those rooted in orthodox Marxism or the new institutional economics) but also readings that maintain that history is pure contingency. I acknowledge, indeed incorporate, the importance of contingencies in the making of past and present, but I argue that the dialectic between culture and ecology, always mediated by human labor, brackets the set of historical possibilities.[8] In keeping with Marx, I understand labor as social, productive, and symbolic activity in the material world: not only does labor produce use and exchange values, but it also produces an agent's sense of self-identity and notions of self-worth in relation to (whether in sympathy with or opposition to) one or more collectivities.[9] Labor, in short, facilitates the co-construction of nature and human nature, a key theme not only of this volume but also of other recent collections.[10]

What differentiates the present collection from others in the genre is its attempt to systematically explore the nature/human nature dialectic within *agrarian* societies (see the editors' introduction, "Agrarian Environments"). This has several implications from my standpoint: first, it restores ecology to agriculture without succumbing to the pitfalls of functionalism inherent in older approaches such as human ecology and agro-ecology that take inspiration from systems theory; second, it examines how (often contradictory) invocations of nature become part of authenticating discourses of domination in agrarian societies; third, it destabilizes the concept of "community" (for instance, *the* village community), not in service of a deconstructive agenda but rather in a constructive effort to show that social units or user groups that are deigned "original" or "natural" in environmental rhetoric (and hence the legitimate stewards of local resources) are often products of conflicted histories and organized around the principle of exclusion rather than inclusion; finally, the chapters in the current volume in various ways pound home the Heideggerian insight that what we identify as reality is ontologically and epistemologically temporal: constituted *by* time and validated *in* time.[11] Each of these four subthemes surfaces in my analysis of agrarian change in Matar subdistrict, although in varying detail.

I propose in this chapter that household—indeed, caste—trajectories of accumulation and decline in Matar have been governed by the interaction between four underlying forces, two dealing with the "nature of work," the other two with "the work of nature." I will elaborate on these phrases shortly. For the time being, suffice it to note that these four mechanisms have interacted to produce a pattern of household mobility that is contingent and multidirectional. The road map for the rest of the chapter is as follows. The first part sets the context by providing basic facts about the physical and social terrain of Matar subdistrict. The second part harnesses ethnographic and survey evidence from two study villages in Matar to illustrate ongoing caste-wise change in the area. The third part develops my propositions about the underlying mechanisms of agrarian change in Matar. The fourth part grounds these claims in historical and contemporary evidence and underscores the four subthemes that I believe are the hallmark of this collection. The fifth part offers a summation of arguments.

A SENSE OF PLACE

In terms of its longitudinal boundary, Gujarat is India's westernmost state. It contains nineteen districts, of which Kheda in central Gujarat is agriculturally the most prosperous. Matar Taluka is an administrative subdivision of Kheda. It forms Kheda's northwestern boundary with adjoining Ahmedabad district. Matar has eighty-two villages and a population of 182,902 (1991 Census of India) dispersed over 225 square miles. Although agriculturally well-off in comparison to other parts of Gujarat, in comparison to its nine sister subdistricts in Kheda, Matar has long been regarded as economically marginal. The soils in Matar are poorer. The fertile sandy loam called *goradu* that is the foundation of Kheda district's agricultural prosperity is confined to a belt in the south and east of Matar, with patches elsewhere. Most of Matar is covered in a low-lying clay loam *(kali jamin)* that sequesters water. This makes the land eminently suitable for rice cultivation, provided there is adequate irrigation. This has not always been the case. For much of history, the viability of agriculture in Matar has depended on the quality of the monsoons. Although annual precipitation in the subdistrict averages thirty inches a year, rainfall records for the period 1876 to 1993 suggest sharp year-to-year fluctuations (the coefficient of variation in rainfall for this period was 0.44).[12]

Cultivation has never been easy in Matar. The lay of the land—a shallow bowl comes to mind—makes drainage in the area sluggish. In years of low rainfall, cultivators in Matar used to face the prospect of a drought; in years of excessive rainfall, the prospect of flooding. Despite the uncertainty of the production environment, not everyone in Matar lived at the edge of subsistence. A number of households from the caste of Kanbi Patels had managed to do reasonably well by securing seasonal irrigation from dug wells, ponds, and rivers, an achievement that was aided by their access to credit from urban traders and moneylenders and, subsequently—after the cotton boom of the 1860s and 1870s—from prospering Kanbi elites known as Patidars.[13] This same financial network appears to have given them the capacity to recover from nature's vicissitudes—droughts, floods, frosts, and epidemics—more quickly than other inhabitants in Matar.

In 1962 Matar's economic fortunes changed dramatically with the extension of three feeder branches from the Mahi Right Bank Canal (MRBC) scheme, a massive surface irrigation project that was underwritten by the central and state governments. Fifty-seven villages in Matar were incorporated into the project's command area. Initially, irrigation was sporadic, but by 1975 63 percent of the net sown area in the beneficiary villages was under irrigation, with the bulk coming from the newly spun web of government canals.[14] The penetration of canal irrigation sparked a resounding change in cropping patterns. Cotton, pulses, and dry cereal varieties were displaced by irrigated cereals. Single cropping expanded to double cropping (and in a few areas even triple cropping). By the 1980s, Matar Taluka had become the "rice bowl" of Gujarat. One visible mark of prosperity was the explosion in the price of irrigated agricultural land, which, by 1995, was selling for up to Rs 60,000 ($1,400) an acre; thirty years ago, an aspiring seller would have been lucky to fetch one-twentieth that price.

The dramatic shift in agronomic conditions and practices has been mirrored by equally remarkable changes in the social landscape. Take, for instance, the Vaghris. One of several so-called untouchable castes in Gujarat,
the Vaghris have a long history of being marginalized. In 1911 British officials in Gujarat accused them of thieving and banditry and branded the entire lot a criminal caste under the Criminal Tribes Act.[15] For several years,
every member of the Vaghri caste over the age of six was required to attend a daily roll call in front of the village headman.[16] In addition, they
were required to provide corvée labor *(veth)* for the entourages of traveling colonial officials. The low standing of the Vaghri was embalmed in
language, particularly in the everyday usage of upper castes.[17] Patels, for
instance, are apt to scold someone behaving in an uncouth manner by saying: "Kem vaghran jevo vyavhar kare chhe?" [Why are you behaving like
a Vaghri woman?]. Vaghris are described by the upper castes as unreliable,
dirty, and foulmouthed. And yet when I was doing fieldwork in Matar,
I was candidly told by several upper-caste informants that "have aainyan
Vaghri no raj chhe" [here, today, Vaghris rule]—an admission that Vaghris
have prospered so much in recent years that their economic prominence
can no longer be denied.

Sabarbhai Tadbda from the revenue village of Shamli is emblematic of
the Vaghris' ascent.[18] As late as 1985, he owned barely half a *bigha* of land
(one bigha in Gujarat equals ⁴⁷/₈₀ of an acre). He now owns seven bighas
dispersed across three villages, has acquired usufruct rights to another
seven, and plies a Tata 407 minitruck, which he purchased in 1994—in
cash—for Rs 315,000 (approximately $7,400). He uses the truck to convey
his tomato harvest to various urban markets, wherever prices are best. In
addition, he lends money on interest and, according to several villagers,
has up to Rs 100,000 ($2,350) in circulation.[19]

The Bharwads, a relatively lowly caste of pastoralists, provide another
example of social churning. As recently as thirty years ago, the Bharwads
occupied the fringes of the village economy. They made a living by tend-
ing the cattle of large Patel landowners, for which they were paid twenty
kilograms of wheat annually by each patron. Today the Bharwads have
achieved economic parity with their patrons and in some cases have supplanted them.

Agro-ecological Change in Central Gujarat 221

As part of my effort to map the emerging economic changes in Matar in the postirrigation era, I conducted census surveys and pursued ethnographic work in three villages. I interviewed a total of 591 households, collecting information on their demographics, a wide range of assets, and their transactions in land, labor, and credit markets. In addition, I was able to conduct open-ended interviews with key informants in at least a third of the eighty-two villages in Matar. Table 1 presents landholding information from the village of Shamli, where I surveyed 353 households and where Bharwads have made striking economic advances in the three decades since irrigation first arrived in Matar subdistrict.

The table reveals Shamli as a caste-heterogeneous village. I counted nineteen different caste groups in Shamli: they are arranged in the table in descending order of their present ritual rank within the caste hierarchy. In this chapter, I limit my attention to four caste groups—the Patels, the Kolis, the Bharwads, and the Vaghris—whose landholding information is consequently highlighted. It is immediately apparent that of these four caste groups, the Bharwads are the only one whose average ownership holdings in 1995 surpassed their holdings in 1965. In fact, every caste group in the village, barring the Bharwads, suffered a decline. I will shortly explain the reason for this anomaly.

But first I want to draw attention to two other telling statistics: the disparities between ownership holdings in 1995 (LAND95) on the one hand and operated and effective holdings (OPLAND95 and EFLAND95) on the other for the four highlighted caste groups. Operated holdings refer to actual land cultivated by a household, taking into account additions or deductions to ownership holdings via tenancy and/or mortgage transactions. Effective holdings denote land that is de facto under the control of a household once its ownership holdings have been adjusted to take stock of mortgage transactions. Table 1 suggests that in the village of Shamli, Patels, Bharwads, and Vaghris have been able to augment their control over land—and hence their economic positions—through tenancy and mortgage transactions, whereas the Kolis have suffered a setback (compare LAND95 to EFLAND95). A closer examination of land transactions in Shamli reveals that in fact the position of Patels as a group has been slipping. Several Patel households have had to sell or mortgage their lands to remain financially solvent. As a compilation of averages, table 1 disguises this fact. The continuing prosperity of two prominent Patel households in Shamli is the

Table 1: Indices of Average Landholding by Caste, 1995, Shamli Main Village (in Acres)

Caste	Average LAND65	Average LANDINH	Average LAND95	Average OPLAND95	Average EFLAND95	Caste Count
Brahmin	12.57	4.94	5.44	5.54	6.31	7
Thakkar	7.35	6.19	5.03	7.35	7.35	1
Patel	**24.76**	**13.52**	**14.79**	**14.71**	**16.30**	**17**
Sadhu	18.27	6.09	6.09	9.28	6.09	2
Rajput	6.38	5.80	5.80	1.26	3.77	4
Suthar/Soni	5.80	0.00	0.00	0.00	0.00	3
Koli/Baraiya	**13.24**	**4.14**	**4.92**	**5.64**	**4.43**	**106**
Prajapati	1.81	1.85	2.34	2.05	2.48	9
Darji	0.00	0.00	0.00	0.06	0.06	5
Bharwad	**10.30**	**8.99**	**11.01**	**11.62**	**11.38**	**55**
Muslim	13.69	8.47	7.31	6.57	6.47	13
Raval	2.74	0.92	0.92	4.88	0.76	11
Vanand	6.19	6.19	3.58	3.58	3.58	3
Od/Pagi	0.00	0.00	0.00	0.36	0.00	4
Vankar	16.21	4.32	5.51	5.23	5.39	20
Rohit	4.54	1.83	2.53	2.25	2.20	53
Vaghri	**2.60**	**1.35**	**1.35**	**2.06**	**1.57**	**21**
Harijan	0.54	0.16	1.38	1.16	0.82	14
Adivasi	0.00	0.00	0.00	1.16	0.00	5
Total households =						353

LAND65: Ownership holdings in 1965 (original undivided household).

LANDINH: Ownership holdings at inheritance (if formal or informal partition occurred).

LAND95: Ownership holdings in 1995 (original or divided household).

OPLAND95: Operated holdings in 1995 (ownership holding + net leased + net mortgaged).

EFLAND95: Land under personal control (ownership holding + net mortgaged).

Caste Count: Total no. of households in a given caste.

Averages are computed as total holdings of each caste group divided by total no. of households in that caste group.

Source: Questionnaire surveys, Shamli, 1994 and 1995.

principal reason for the improvement in the average operated and effective landholdings of the Patel caste.

Why have Bharwads prospered so noticeably? According to table 1, Bharwads as a group owned an average of 10.3 acres of land in 1965, a figure that by 1995 had crept up to 11 acres despite some fragmentation of holdings at inheritance (contrast LAND65, LANDINH, and LAND95). What table 1 does not display is that in 1965 the land that Bharwads claimed as their own was in fact vested under the de jure ownership of a government-endorsed fodder cooperative called the Gopalak Mandali and produced primarily grass. The Bharwads surreptitiously divided the cooperative's land into individual ownership parcels in contravention of the conditions under which the land had been transferred to them by the state. By 1995 the Bharwads were able to augment their de facto privatized holdings by purchasing and acquiring control of additional village lands. Moreover, they were able to convert their previously untended "private" holdings that yielded seasonal bursts of grass into irrigated cereal production: rice in the *kharif* (monsoon), wheat in *rabi* (winter).

The contrast between the economic performance of Patels and Bharwads since the arrival of canal irrigation in Matar is underscored by an examination of land purchase and sale records maintained at the Subdistrict Registrar's Office in Matar town. It is worth noting here that Patels as a norm attempt to keep land sales endogamous: Patel households who want to sell holdings face tremendous peer pressure to sell to fellow caste members. Although ordinarily fractious in their dealings, Patels are acutely aware of the need for solidarity in matters of caste dominance *(qum satta)* and therefore the importance of retaining control over land in a primarily agrarian economy.[20] Yet the subdistrict registrar's records show that in 1994 four Patel households in Shamli sold 58.3 acres to Bharwad families and a paltry 3.6 acres to a fellow Patel.[21] This pattern of transactions reveals first of all the buying power of Bharwads, who, in this case, paid up to Rs 60,000 (roughly $1,400) an acre for the land they acquired; second, it indicates that solvent Patel households in Shamli today can neither match the purchasing power of Bharwads nor effectively deploy the threat of moral condemnation as a way of internalizing land sales.

A study of mortgage transactions in the villages of Shamli and Astha confirms the emerging picture. The practice of mortgaging land, known locally as *giro pratha,* has a long history in Kheda district.[22] In return for a loan, the landowner (or mortgagor) temporarily cedes control of his land to the lender (mortgagee). The length of mortgage agreements varies between a year to seven years, with explicit provision for contingent renewal on an annual basis in the case of some one-year contracts. There seems to be wide acceptance of a principle that if a loan is not repaid by the end of seven years, it is legitimate for the lender to ask the borrower for a transfer of title.[23] Thus, depending on the nature of the mortgage transaction, it resembles either a straightforward fixed-rate tenancy contract or, more commonly, an interlocked exchange involving credit for land.

Although the motivations of landowners for out-mortgaging land vary, the transaction itself has increasingly come to symbolize erosion in a household's financial solvency—primarily because temporary transfers of use rights to land in return for a loan (the yield from the mortgaged land representing interest on the loan) have increasingly dissolved into permanent transfers. In short, the mortgage transaction has become an indicator of household mobility rooted in local cultural understandings of economic well-being.

The story of mortgages in Astha and Shamli (summarized in tables 2 and 3) is revealing. Table 2 documents the incidence of households by caste group that have transferred part or all of their land on usufructuary mortgage (are mortgagors), whereas table 3 reports the incidence of households that have retained part or all of somebody's land (are mortgagees) as loan collateral. Jointly, they expose the emerging patterns of inequality in the case villages. Whereas the average incidence of out-mortgaging among reporting households is 26 percent in Astha and 20 percent in Shamli, the incidence of out-mortgaging by households of the Koli/Baraiya caste is noticeably higher than the village average: 44 percent in Astha and 36 percent in Shamli. In conjunction with the evidence presented in table 1, these figures bolster the image of Koli/Baraiyas as a caste sliding into semiproletarianism (but as I will show in the fourth part, the trend toward full proletarianization has been arrested by nature's unscripted interventions).

The pattern of mortgage transactions also establishes the strong economic position of Bharwads relative to other caste groups: not a single Bharwad household in either Astha or Shamli appears as a mortgagor, indicating their lack of compulsion to obtain credit by offering land as collateral. By contrast, a few Patels do appear as mortgagors—2 out of 22 households in Astha, and 5 out of 17 households in Shamli. Strikingly, Vaghris

Table 2: Land Mortgagors in Astha and Shamli by Caste Group, 1994–1995

Caste Group	ASTHA			SHAMLI		
	Mortgagors (No. of Hhs) (1)	Total Reporting Hhs (No.) (2)	Percent of Mortgagors (3) = (1) ÷ (2)	Mortgagors (No. of Hhs) (4)	Total Reporting Hhs (No.) (5)	Percent of Mortgagors (6) = (4) ÷ (5)
Brahmin	0	6	0	1	7	14.29
Thakkar	0	4	0	0	3	0
Patel	**2**	**22**	**9.09**	**5**	**17**	**29.41**
Rajput	0	1	0	1	4	25.00
Sadhu	1	3	33.33	0	2	0
Panchal	4	26	15.38	1	2	50.00
Koli/Baraiya	**27**	**62**	**43.55**	**38**	**105**	**36.19**
Prajapati	0	4	0	0	9	0
Lohar	0	0	0	0	1	0
Darji	0	0	0	0	5	0

	49	190	25.79	71	353	20.11
Bharwad	0	**10**	0	1	**55**	0
Muslim	0	0	0	4	13	30.77
Raval	0	0	0	1	11	9.00
Vanand	2	5	40.00	0	3	0
Dhobi	1	4	25.00	0	0	0
Pagi	1	4	25.00	0	3	0
Vankar	0	0	0	2	23	13.04
Nayak	1	2	50.00	0	0	0
Rohit	6	18	33.33	11	53	20.75
Adivasi	0	0	0	0	3	0
Vaghri	0	1	0	0	**21**	0
Harijan (Bhangi)	4	18	22.22	6	14	42.86
Total	49	190	25.79	71	353	20.11

Source: Fieldwork data, 1994–1995; caste groups are arranged in descending order by ritual rank.

Table 3: Land Mortgagees in Astha and Shamli by Caste Group, 1994–1995

Caste Group	ASTHA			SHAMLI		
	Mortgagees (No. of Hhs) (1)	Total Reporting Hhs (No.) (2)	Percent of Mortgagees (3) = (1) ÷ (2)	Mortgagees (No. of Hhs) (4)	Total Reporting Hhs (No.) (5)	Percent of Mortgagees (6) = (4) ÷ (5)
Brahmin	0	6	0	2	7	28.57
Thakkar	1	4	25.00	0	3	0
Patel	9	22	**40.91**	7	17	**41.12**
Rajput	0	1	0	0	4	0
Sadhu	0	3	0	2	2	100.00
Panchal	5	26	19.23	2	2	100.00
Koli/Baraiya	**10**	62	16.12	**12**	**108**	**11.43**
Prajapati	1	4	25.00	3	9	33.33
Lohar	0	0	0	1	1	100.00
Darji	0	0	0	1	5	20.00

Bharwad	2	10	20.00	10	55	18.18
Muslim	0	0	0	0	13	0
Raval	0	0	0	0	11	0
Vanand	0	5	0	0	3	0
Dhobi	0	4	0	0	0	0
Pagi	0	4	0	0	2	0
Vankar	0	0	0	1	23	4.35
Nayak	0	2	0	0	0	0
Rohit	2	18	11.11	0	53	0
Adivasi	0	0	0	0	3	0
Vaghri	0	1	0	3	21	14.29
Harijan (Bhangi)	0	18	0	0	14	0
Total	23	190	12.11	44	353	12.46

Source: Fieldwork data, 1994–1995; caste groups are arranged in descending order by ritual rank.

are entirely absent from the list of mortgagors; more significantly, three Vaghri households (including Sabarbhai Tabda's) have kept land on mortgage—indicating their strong financial position. Overall, the evidence in tables 1 through 3 confirms the image of a society in flux where previously marginalized groups are beginning to challenge the economic supremacy of the historically dominant Patels. Key informant interviews in various other villages of Matar subdistrict validate this conclusion.

MECHANISMS OF CHANGE

Can we think systematically about the emerging pattern of agrarian change in Matar, where yesterday's subalterns are becoming today's superiors? I suggest we can. And I propose that there are four mechanisms at work. The first is the ability of households to manage production risk and consumption demands by securing access to a diversified space economy.[24] But this mechanism, which has a positive effect on a household's ability to accumulate, has to be juxtaposed against a second—countervailing—mechanism: namely, the desire of households to position themselves in social space by shunning certain forms of labor. This is normally costly and functions as a drain on accumulated surpluses. I label these two opposing strategies "the nature of work." The third mechanism is simply the subsidy that nature provides to various livelihood activities. The subsidy is not uniform: some ecological spaces may be more valuable than others from an economic standpoint. But this may rapidly change. And here we get to the fourth mechanism: nature's unpredictability. Human attempts to transcend nature's limits—through canal irrigation and biochemical technologies, for instance—may have unanticipated consequences. I call the third and fourth mechanisms the "work of nature."

I am going to say a little more about each of these four mechanisms. But notice what I am offering: a model of agrarian change that combines loose determinism with contingency. History is neither the predictable expression of unseen behavioral or structural "factors" nor mere accident. Instead, my argument claims that the patterns we observe in Matar are an outcome of how four broad processes have dialectically interacted over time. And although my claims in this essay are place specific, I am confident that they contain analytical leverage for understanding agrarian transformations elsewhere, particularly in other parts of south and Southeast Asia.

The four mechanisms I have proposed merit elaboration:

Mechanism 1 Trajectories of accumulation or decline in Matar since the 1830s can be understood only in the context of local, regional, and global geographies of work—specifically, the ability of households to secure access to labor surpluses at key historical moments from various niches and corners of a differentiated space economy. I contend that these surpluses have been obtained in three ways: by livelihood diversification, through credit networks, and via access to kin or community labor that can be "superexploited." Notice that each of these strategies involves the transfer of surpluses between a part of the space economy that is robust and another that is ailing. The transfer can be either in the form of money (strategy one), in the form of credit (strategy two), or in the form of labor (strategy three). But regardless of form, each of these modes of surplus acquisition has allowed households to manage, to varying degrees, the production and consumption risks they have faced at various times. I am suggesting (and by no means originally) that the ability to negotiate risk through the circulation of labor and capital is a prerequisite for accumulation in a semi- or fully capitalist agrarian system.[25]

Mechanism 2 Work—or the act of laboring—is a material *and* symbolic activity. Work is not only the way each of us makes a living but also the way we create ourselves in relation to others through the meanings invested in forms of work. This is not a new insight. It is borrowed from Marx and has been extensively developed by Hegelian Marxists and members of the Frankfurt School.[26] The "take-home" point is that while work of some kind is a practical necessity, it is also a powerful instrument for establishing social distance. When we reconstruct the history of local production relations in Matar from the mid–nineteenth century onward, we can trace the gradual emergence of a topology of work, where some forms of work came to be regarded as superior to others. These forms became sources of social distinction. But this quest for distinction within the labor process was expensive. It weakened the armor of Patel domination. Subordinate groups were eventually able to exploit this vulnerability to undermine the Patels' preeminence.

Mechanism 3 The economic success or failure of households in Matar has always depended on the way they have been able to use ecological spaces and harness nature's subsidy. This is what I mean when I say that nature performs work. Here a brief digression is necessary to clarify where this paper stands in relation to recent debates about the concept of "nature," a term that Raymond Williams in his quirky but insightful volume *Keywords* recognizes as one of the most slippery in the English language. In a recent article, David Demeritt identifies five distinct philosophical approaches to nature. His taxonomy includes "common sense realism" (the foundationalist view of nature as objective reality whose structure can be validated through propositional statements); "social object constructivism" (nature as ontologically dual, consisting of a social reality that is cultural and a material reality that is not of human making); "social institutional constructivism" (nature is pregiven but known through epistemological frameworks that evolve historically); "neo-Kantian constructivism" (the superidealist view that reality is representation and as such nature is what discourse makes of it); and "artifactual constructivism" (nature has ontological existence, but its reality emerges through material practices).[27]

My philosophical position lies closest to "artifactual constructivism," with two modest qualifications. First, unlike its proponents who speak of the co-construction of nature (what we recognize as "nature" is a collaborative production of human and nonhuman actors) primarily within the context of laboratory experiments and scientific knowing,[28] I want to draw attention to nature as a biophysical reality that is negotiated, transformed, and culturally deployed (in essentializing and dominating ways) in the everyday routines of agrarian life. Second, this biophysical reality has to be regarded as "referentially detachable" and "transfactually efficacious"[29] — in other words, manual or mental interventions in nature in order to derive use values from it require human beings to *believe* that their labor is performed on an entity that is *practically speaking* external to them (i.e., "existentially intransitive") and whose response is governed by certain empirical regularities. Asserting the quasi independence of nature *during moments of (manual and mental) production* recognizes that social relations are mediated through practices that invariably rely on the "taken-for-granted" (or structural) character of reality.

Mechanism 4 There is another important caveat to the equation: nature's work, which I have proposed as one of the key mechanisms of agrarian change in Matar subdistrict, has to be supplemented with natural agency of a different kind—nature's unpredictability in the face of human interventions. Here, I want to draw attention to the fact that nature sets definite limits on accumulation. Not just the kinds of limits agro-food geographers talk about, where nature hinders capitalist penetrations of agriculture by creating a disjuncture between "necessary labor time" and "production time";[30] those limits, while clearly important, have to be juxtaposed against the stochastic character of our interactions with nature. Although we may try, we cannot regulate nature's services or subsidies with algorithmic certainty. Nature's agency, in short, has a simultaneously enabling and disabling character—a point that I hope will be abundantly clear when we examine the experience of canal irrigation in Matar.[31]

But even in the abstract, the dual character of nature's agency is not difficult to grasp. Take agriculture. As Ted Benton points out, labor processes in agriculture have an "ecoregulatory" character rather than the strictly "transformative" character of industrial production (1996, 160–63). This has several implications, consistent with a modified "artifactual constructivism":

1. Agricultural production combines transformative moments when labor is applied to nature with nature's work in the form of "transfactually efficacious" organic processes that encompass soil properties, species interactions, and the gestational logic of cultivated species (whether crops, livestock, poultry, or fish).

2. Nature within agrarian environments figures "both as *conditions* of the labor process, *and* as *subjects* of labor" (Benton 1996, 161).

3. Regardless of the technical organization of agriculture, there will always be some aspects of nature that will remain "relatively impervious to intentional manipulation, and [may] in some respects . . . [be] absolutely nonmanipulable" (Benton 1996, 161).

In short, the ecoregulatory character of agriculture builds a fundamental uncertainty into this human enterprise.

TOWARD GROUNDED THEORY

The narrative I am about to weave will demonstrate how the four mechanisms I just outlined have informed the trajectory of agrarian change in Matar. As a first step, let's briefly step back in history to understand how the Kanbi Patels in fact became the dominant caste in Matar and the rest of central Gujarat. When the British first acquired control of central Gujarat between 1803 and 1817, there were two major communities in the region: the Kolis and the Kanbis.[32] Neither the Kolis nor the Kanbis were as yet castes—and here I am applying the accepted usage of the term "caste" as an endogamous group with a specific ontology, a specific claim to social rank, and a specific set of ritual practices.

Early colonial accounts instead suggest the Kolis and Kanbis were socially differentiated on the basis of habitation in the area. Kolis are presented as prior inhabitants of the region, and Bishop Heber in his travelogue speculates that they might be ritually degraded Rajputs (descendants of intermarriage between Rajput refugees and local tribals, whose caste status was therefore liminal).[33] Kanbis are viewed as later migrants, and their caste status is equally ambiguous. David Pocock, for instance, drawing on ethnographic work in rural Kheda in the 1950s and 1960s, describes meeting Brahmins who regarded Kanbi Patels as little better than upstart Shudras (the lowest tier in the stylized fourfold caste hierarchy). Some Kanbis even today concede their modest origins, but they do so primarily to demonstrate how far they have risen in the economic and ritual order of central Gujarat. Other Kanbis are less forthcoming: when quizzed about origins, they blankly declare ancestral ties to Rajputs and ritual parity with Vanias (next to Brahmins, the two upper-caste groups in Gujarat).

Alice Clark makes the persuasive case that when colonial rule first arrived in central Gujarat, the economic and ritual positions of ordinary Kolis and Kanbis were roughly similar: not only were they indistinguishable in dress and disposition to early travelers and colonial officials, but there is in fact substantial evidence of intermarriage between Koli and Kanbi plebians (particularly of Kanbi men marrying Koli girls). Clark concludes that the "caste" identities of Kolis and Kanbis were as at best porous, at worst inchoate.[34]

This discussion of economic and ritual ancestry provides a simple illustration of not only the unnatural (i.e., historically transitive) character of groups—including those we call "communities"—but also, as I now show, the links between power and group formation. Although caste identities were murky, class divisions were apparently less so: colonial documents suggest that both Kolis and Kanbis (the latter more so) were economically and culturally stratified into a class of ordinary cultivators and a class of aristocrats who had risen to positions of considerable authority in the Muslim and Maratha administrations that preceded British rule.

Colonial officials initially adopted an impartial attitude toward both groups. But around the 1820s, there was a historic shift. District-level colonial officials—those entrusted with everyday administration—appear to have concluded that the Kanbis were the backbone of agriculture in Gujarat and should therefore be favored.[35] That Kanbis were sedentary cultivators, while the majority of Kolis preferred shifting cultivation, seems to have been a significant factor in galvanizing this attitudinal shift—after all, in those early days of colonialism, the revenue imperative loomed largest in the minds of district officers. Many of these officers were, moreover, groomed to the virtues of private property, and a group of stationary cultivators must have sat far better with this ideal than a group of moving cultivators. Kanbi elites also did their part to tilt colonial policies in their favor by monopolizing channels of representation.[36]

By the 1830s and 1840s, the economic and political influence of Kolis—commoners as well as aristocrats—was in steep decline. This was precisely the time when Kanbis extended their power over central Gujarat's economy and society. Despite sharp internal class divisions, the ideological cement of an emerging caste identity seems to have been sufficiently adhesive to bind the Kanbis into a corporate community. This meant that they were able to handle the political opportunities and challenges of the times as a collective. They petitioned colonial officers and protested colonial policies as a group. They employed group suasion—and sometimes strong-arm tactics—to subdue local rivals, particularly Kolis.[37] Most importantly, when it came to cultivation, they aided each other financially. Kanbis, moreover, were able to tap into spatial networks of credit whose linchpins were Jain and Thakkar Vanias and affluent Patidar (wealthy Kanbi) families from the fertile heart of Kheda district.[38] Ready access to credit allowed Kanbis to invest in land improvements, weather agronomic risks, and practice a better form of agriculture than other groups.[39] So linked were credit and social prestige that the two have become semiotically entangled: hence the Gujarati word *abroo,* which Patels frequently invoke as a way of describing reputation, honor, or family prestige, also means "creditworthiness."[40] The ability of the Kanbis to mobilize credit for economic and social mobility is an early example of mechanism 1 at work: risk management, hence the possibility of capital accumulation, by movement of surpluses within a space economy with uneven patterns of development and asynchronic growth cycles.[41]

I also want to suggest that initial class differentiation among the Kanbis was indispensable to the construction of a distinct caste identity, and that an emergent caste identity was, in turn, instrumental for future class accumulation.[42] Kanbi aristocrats (the Patidars) and ordinary Kanbis associated strategically, but unequally, in the making of the "Patel" caste. The term "Patel," which originally applied to the village official in charge of tax collection and law and order, was now adopted wholesale by Kanbis as a mark of their community's rising prestige. The formal declaration of this amalgamation came with the 1931 Census—that quintessential project of modernity: a state trying to make its subjects "legible"—when even the lowliest of Kanbis insisted on recording their caste as "Patidar."[43]

In fact, the economic ascent of less well-to-do Kanbi Patels in Kheda was catalyzed by a "Conradian moment": a cataclysmic famine in 1899 to 1900 that shaped at least two generations of agrarian change. Between 1899 and 1918, there were nine years of heavy or complete crop failure in Matar as a result of inadequate rainfall. The famine of 1899 to 1900, inscribed in folklore as the *chhapaniyu dukal,* was particularly devastating.[44] The subdistrict's population plunged from 79,000 people in 1891 to 62,000 by 1901. The famine also wreaked havoc on the cattle population: milch cattle, draft cattle, and young stock perished in alarming numbers. Several desperate cultivators sold their agricultural implements to survive. Food camps in the district were inundated by hunger-stricken families. Patels were among them, but because of their superior resources, they were able to cope with, and recover from, these two calamitous decades less traumatically than members of other caste groups, whose control over production means and whose access to urban social networks were far more tenuous.[45]

It was in fact in these two decades that the first wave of Patels migrated to East Africa and struck the roots of a diaspora that was to later become extraordinarily powerful. Part of the wealth accumulated abroad, first in salaried and pensioned colonial jobs, later in trade and commerce, was repatriated by Patels to their places of origin in Kheda, where it served to further concentrate assets in the hands of their community. In short, through the spatial—indeed, global—diversification of work, the Patels were not only able to ride out a sequence of natural calamities but emerge even stronger at the end of it. The events described illustrate the importance of mechanisms 1 and 4—risk management through livelihood diversification and nature's semiautonomous agency—in shaping agrarian change in Matar.

But while livelihood diversification was a centripetal force that aided the Patels' accumulation of wealth, their urge to display their status acted as a centrifugal force that strained at their assets. Strategies of work, specifically the resources that became available or were expended as a result, were to become the boon *and* the bane of Patels.

Colonial records suggest that the Kanbi Patels were an exceptionally industrious and enterprising group of cultivators.[46] The fertile sandy loam that is Kheda district's claim to agricultural fame is even today acknowledged to be the outcome of several hundred years of meticulous cultivation practices: nature painstakingly reworked. Even the low-lying clay loams in Matar that sequester water and are so conducive to paddy cultivation are part of a deliberately created landscape. During fieldwork in Matar in 1994 and 1995, I frequently heard the refrain, "Patelón ni kheti ghaní saarí chhe" [Patels practice superior cultivation] or "Je gaam ma Patel ni vasti, te gaam samrudhh" [villages where Patels reside are likely to be prosperous]. In an exaggerated vein, some Patels add that even the presence of a single Patel household is sufficient to alter the complexion of a village for the better. When asked how they have managed to become so powerful in Kheda, Patel *agevans* (community leaders) invariably invoke a discourse of superior genes and "natural qualities" *(prakritik gun)*. Evidently the Patels are just as skilled at cultivating a self-ideal as they are at cultivating nature. In fact, hierarchy and the idea of self-refinement seem to have become prominent aspects of Patel identity from a very early stage of their coagulation as a caste.[47] Patels are fastidious about their food and their attire. And although most villagers in Matar defecate in fields, a Patel household (no matter how poor) will insist on maintaining a latrine or toilet; not necessarily because it's more hygienic—it probably isn't given the lack of underground drainage—but because it's considered a mark of civilization.

Similarly, Patels have endeavored to acquire distinction through perceived refinements in the way they deploy their labor. Very early on—by the 1860s or 1870s—Kanbi Patels started disengaging from wage labor. A struggling Patel male preferred to be a sharecropper receiving some proportion of the harvest than a wage laborer. Similarly, if he could muster money for rent, he preferred to be a tenant rather than a sharecropper.[48] These were basically attempts to bracket the manner in which labor was valorized. Second, Patel cultivators began to shun physical labor in favor of supervisory or mental labor that could be performed, more or less, at will. Third, Patels withdrew women's work not only from the commodity circuit—which in any case was rare—but from the public gaze as well. Any agricultural work that a woman performed was now confined to the private domain of the family residence.[49] These are illustrations of mechanism 2.

The important point is that each of these work strategies was costly. The expense took the form of an opportunity cost—either forgone labor income or agricultural output—that imposed a financial burden on Patel farmers. But it counted as distinction *precisely* because it was costly and few in a basically subsistence economy could afford to be choosy about how or where they deployed their labor.

The past three decades have witnessed yet another modification in the nature of work. The change is generational. Younger-generation Patels prefer to conduct agricultural operations from a distance. Like their fathers and grandfathers, they exhibit a preference for supervisory labor but unlike them are rarely willing to spend their days overseeing agricultural operations in the fields—instead, supervision consists of ordering intermediaries from the comfort of the village, delegating tasks, or occasionally darting by motorcycle to monitor ongoing operations. Spending entire days in the field is now regarded with disdain.

This new form of work distinction has also come at a cost. It has altered the architecture of the local labor market because young Patel farmers now like to delegate agricultural work on piece rates, where they simply pay a team of workers a lump sum for completing a task. Because a piecework regime requires minimal supervision, it suits the younger Patels with their newly acquired taste for leisure. But piecework may make poor economic sense. On one occasion, an older Patel farmer angrily denounced the young son of a wealthy Patel landowner who had agreed to pay a group of workers Rs 250 to spread organic manure over roughly two acres of land.[50] He was aggrieved because according to him, the same task, with supervision, could have been accomplished at a third of the labor cost, for about Rs 80. In his opinion, decisions like the one made by the young farmer were irresponsible because they encouraged laborers to raise their demands.

To fully appreciate the older farmer's concerns, we must place them in the context of other emerging developments in Matar. Canal irrigation, since its arrival in the early 1960s, has had some unexpected effects. The expansion in the area under paddy by roughly forty thousand acres since 1965 has dramatically increased the supply of dry fodder, because paddy straw—a by-product of cultivation—is a nutritive food for livestock.[51] Meanwhile the explosion of grass along the irrigation minors and subminors that snake through the landscape has created new grazing areas for open-grazed live-stock, and a steadier supply of green fodder for stall-fed milk animals like buffalo and hybrid cows. Ecological spaces that were earlier economically marginal have suddenly become valuable.

This unanticipated improvement in dry and green fodder availability—an illustration of nature's work and agency—has encouraged the spread of village-level dairy cooperatives in Matar and sparked rapid growth in the livestock sector. The number of village dairies in Matar escalated from thirty-two in 1965 to seventy by 1990, and membership in them increased fivefold.[52] Suddenly the milk economy has become very important in Matar and created an alternative to wage labor for traditionally subordinate groups such as the Kolis and Bharwads.

This *decrease in the elasticity of labor supply* has to be juxtaposed with the *rise in labor demand* that has accompanied the massive expansion in paddy acreage—one of few commercial crops that has high tolerance to waterlogging and which cultivators in Matar could therefore safely adopt given the prevalence of water-retaining clay loam soils and the area's poor natural drainage. Here again we find evidence of nature's concurrently enabling and constraining aspects. The net effect of expansion in paddy cultivation has been a significant augmentation in the bargaining power of rural working classes.[53] In other parts of the world that have experienced similar processes of intensification, employers have responded to the increase in labor's bargaining power by resorting to mechanization of agricultural operations. Mechanization can ease temporal constraints that become more binding with the increase in cropping intensity, but more importantly for employers, by easing dependence on labor, mechanization can serve as a worker-disciplining device.

The extent of labor savings induced by tractorization—the primary mechanical innovation adopted in India and in Matar subdistrict—is ambivalent.[54] Tractors, in fact, appear to have complemented rather than displaced labor: the only agricultural operation where tractors directly compete with labor is in the carting of fertilizers to the field and harvested crops away from it, but even in these operations, substitution appears to have been minimal. In any case, the existing literature on mechanized agriculture suggests that labor displacement is most pronounced in the case of combine-harvesters, which intrude on two of the most labor-intensive crop operations: harvesting and threshing. One reliable estimate of labor requirements for harvesting and threshing of HYV (high-yielding variety) paddy places the figure at 32 percent of total person-days spent in crop-related operations. The proportion for harvesting and threshing of HYV wheat is similar, although absolute labor absorption in wheat is considerably smaller. Hence any use of combine-harvesters can be expected to have a direct negative effect on labor demand. Whereas combine-harvesters are used to harvest both paddy and wheat in Punjab and Haryana (the two states that are emblematic of green revolution agriculture in India), in Matar they are used only to harvest wheat. Why? Here once more we are forced to take cognizance of nature's agency.

Two large farmers in Astha, one of my three study villages, informed me of unsatisfactory experiments several years ago to remove paddy by harvesters. The attempts failed because the harvester could not operate easily in the viscous, waterlogged soil. Moreover, mechanical harvesting damages paddy stalks and renders them unfit for use by cattle. This is of great concern to cultivators because not only are paddy stalks the primary source of stored fodder for milch cattle, but in addition their sale by trailer loads in adjacent fodder-deficit areas provides substantial supplementary income. In short, the opportunity cost of mechanical harvesting is perceived to be too high for paddy. To summarize, while it is abundantly clear that employers are interested in cutting labor costs and disciplining local laborers by mechanizing labor-intensive agricultural operations, they have so far experienced only limited success in this endeavor.

Earlier in this chapter, I documented the meteoric rise of Bharwads in the village of Shamli and how they have begun to threaten Patel supremacy there. In the village of Astha, where the population of Bharwads is smaller, the Patels remain dominant but feel besieged. They harbor great animosity toward Bharwads and accuse them of profiteering at the expense of other cultivators, by wantonly allowing their cattle to graze among ripening crops. Although dairying provided the spark, there are several notable elements in the Bharwads' spectacular rise to affluence: among them, as I mentioned earlier, the illegal privatization of community grazing lands originally granted to Bharwad fodder cooperatives under an affirmative action program. The Bharwads, intent on becoming substantial cultivators and landowners, have pumped their large and steady cash flows from milk sales into land purchases and mortgage acquisitions, moneylending, and the creation of intracaste credit cooperatives, which provide loans at low, sometimes zero, rates of interest. The rise of the Bharwads illustrates three of the four mechanisms of agrarian change that I have proposed: mechanism 1, risk management through credit and livelihood diversification, and mechanisms 3 and 4, nature's subsidy and nature's unpredictability, have also been significant factors in the Bharwad's economic ascent. The attempt to transform nature through canal irrigation has made the work spaces where Bharwads traditionally toiled—uncultivated lands and fallows—unexpectedly productive.

Patels, of course, claim that the dairy has had a damaging impact on agriculture by aggravating a labor shortage. Laboring households, they lament, now prefer to maintain a buffalo or two and live off that income rather than earn a proper wage income. One irate Patel denounced livestock income as "aalas ni aavak" [idle income] and declared that it had robbed laborers of their desire to work. But Patels have other reasons to worry, as well. Matar, as mentioned, has clay loam soils. These soils, furthermore, are sodic-alkaline: they have pH values between 8 and 9. Large portions of the subdistrict are prone to water retention and salinization because of the bowl-like topography. A cropping regime that favors hydric crops, heavy reliance on biochemicals, misuse of water (since canal irrigation is available for a flat per acre rate), and shorter and shorter fallows has begun to generate chronic waterlogging. Far worse, it has encouraged a kind of obdurate salt infestation where the salt rises to the surface with capillary action in such quantities that when it dries, it leaves behind a salt efflorescence that makes cultivation either impossible or hopelessly expensive (since the salt first has to be flushed away). Tied to this is a curious tale of globalization, which, in the guise of economic liberalization, arrived at the grassroots in 1992. In that year, the government of India, succumbing to pressure from the IMF, deregulated the prices of all fertilizers except urea. This led farmers to substitute urea for other fertilizers. However, it is now becoming increasingly apparent to farmers and agricultural extension agents that the overapplication of urea has aggravated the problem of salinization. So nature has once again upset human endeavors. And Patels stand to lose the most, since they tend to be the largest landowners in Matar.

Finally, a word about the Vaghris. You may remember them as a caste group, who throughout history were marginalized but in recent years have prospered greatly. The Vaghris used to inhabit the margins of society and make a living off lands that lay at the margins of cultivation. Fishing, hunting, and moonshining were, until recently, their primary sources of livelihood. The Vaghris have relied on family and kin labor to engineer their economic ascent. They have begun to lease lands for tomato and melon cultivation, often paying stupendous rents. Moreover, tomato and melon are crops that require heavy inputs of fertilizers, pesticides, and purchased tube well water. Vaghris defray the enormous risk that accompanies their cultivation of these cash crops by minimizing their use of hired labor. They depend almost entirely on family labor and exchange labor from members of their community: in short, risk management via mobilization of labor surpluses across space. Sometimes the wife's parents travel from their villages to help out in cultivation. The case of the Vaghris illustrates how the "superexploitation" of family or kin labor can oil the wheels of accumulation. The importance of this work strategy can be traced to Marx's distinction between "labor" and "labor power." Labor power—a certain number of hours of work—can be purchased; labor—the effort, commitment, and creativity that a worker puts into his or her creation—cannot be purchased. Surplus accumulation or productivity hinges on labor, which has to be either extracted or volunteered. It isn't difficult to see why it may be far easier to obtain labor effort from members of kith and kin than a hired worker. Vaghris have harnessed this principle to great effect.

CONCLUSION

I want to sum up by returning to the polarization thesis. I think the polarization thesis and its variants *have* supplied a powerful framework for analyzing inequality. The idea that the rich get richer is intuitively appealing. It is even ingrained in common sense. But the thesis itself may be an example of an "ecological fallacy" masquerading as general theory. Even if the link between initial and future assets holds at the level of regional or national aggregates such as classes or income groups, it is not necessary that the same characteristic apply to subgroups and individuals who make up those aggregates. The link has to be confirmed empirically. The history of agrarian change in Matar subdistrict suggests that the polarization thesis may be incomplete rather than flat-out wrong. After all, the Patels *were* able to consolidate their wealth for more than a hundred years—and readier access to credit, one of the salient factors identified by advocates of the polarization thesis, *was* a key element in their success. But then came their derailment. And significantly, it has come at a historical juncture when it was least expected: after the advent of canal irrigation and green revolution agriculture. Equally astonishing was the rapid economic ascent of two groups, the Bharwads and Vaghris, who had been historically marginalized. The polarization thesis and its variants simply cannot explain *this* kind of social churning.

Why not? I argue in this chapter that explanations of agrarian change—and more generally "agrarian environments"—have to be firmly rooted in time and space, and in *uses* of time and space (both physical and metaphoric) by social actors. These uses emerge from a "logic of practice" whose concrete expression is a set of "generative dispositions."[55] Therefore, the first step in mapping "agrarian environments" is to make intelligible the social conditions of their production. My narrative summarizes this practical logic as the "nature of work." The second step is to theorize the workings of nature, keeping in mind that within agrarian settings, nature is always encountered through labor via some set of "ecoregulatory" technologies. Drawing on the realist philosophies of Roy Bhaskar and Ted Benton, I claim that nature, while always produced, can never be entirely subsumed by human knowing or algorithmically scripted by human reason. In short, nature is neither an autonomous entity nor a mere imagining: it is a bit of both. The third methodological step is to chart the society-nature dialectic: the complex interaction between historically endowed human dispositions and nature's agency that manufactures an observed pattern of social striation and flux. The state, in its myriad forms, is a looming presence in any such narrative.

Anchored in this implicit analytic framework, this chapter has proposed that there are four mechanisms that have propelled agrarian change in Matar subdistrict. Two of these—first, the ability of households to manage risk through participation in a wider-space economy, and second, the quest for distinction through forms of labor deployment—have to do with material practices that I term the "nature of work." The remaining two processes deal with the "work of nature"—on the one hand, the resources nature provides for generating agricultural livelihoods, and on the other hand, the often unpredictable limits it places on accumulation. These dual aspects of nature's agency can work at cross-purposes (they can be simultaneously enabling and disabling) just as the dual aspects of a work geography can exert countervailing pressures on household accumulation. The balance of these forces can not be predicted *ex ante*. They can only be interpreted *ex post* through a critical historical geography. Thus, in Matar, unexpected conjunctures of nature and culture have eroded the Kanbi Patels' long domination of rural society and allowed previously subordinate caste groups to rapidly augment their economic status.

NOTES

This chapter has benefited from the keen insights of Navroz Dubash, Michael Goldman, Martin Olson, Jeff Romm, Rachel Schurman, the editors, co-contributors, and the unminced remarks of two anonymous referees. Thank you all. Financial support to nurture these ideas has come from the Population Council, which gave a generous dissertation-writing award, and the Izaak Walton Killam Foundation of Canada, which provided an equally generous postdoctoral fellowship.

1 Notably Goodman et al. 1987. Also see Goodman and Redclift 1991.

2 There is a broad consensus on this among historians of central Gujarat. See Pocock 1972; Clark 1979; Bates 1981; Hardiman 1981; Chua 1986; den Tuinder 1992.

3 Kautsky [1899] 1990; Lenin [1907] 1974.

4 For a concise description of the "theory of inevitable polarization" see Netting 1993, 214–21.

5 The principal advocates of this position are Mann and Dickinson (1978); Mann (1990); Goodman et al. (1987); and Goodman and Redclift (1991).

6 Cf. Kautsky [1899] 1990; Chayanov [1966] 1986; A. Sen 1966; Ellis 1988, chap. 6.

7 The theory of "functional dualism" owes to Karl Kautsky ([1899] 1990). Applications of it may be found in de Janvry 1981 and in Deere and de Janvry 1981.

8 "What we experience and theorize as nature and as culture are transformed by our work. All we touch and therefore know, including our organic and our social bodies, is made possible for us through labour. Therefore, culture does not dominate nature, nor is nature an enemy. The dialectic must not be made into a dynamic of growing domination" (Haraway 1991, 10).

9 Marx [1844] 1988, esp. 69–84, 141–70. Also see Donham 1990.

10 Haraway 1991; Benton 1996; Peet and Watts 1996; Braun and Castree 1998.

11 This reading of Heidegger follows from "critical realist" philosophies of knowledge, especially Roy Bhaskar (1989, 1993). It recognizes the historicity of human knowing and the temporality of reality without reducing reality to representation: in other words, it concedes epistemological transitivism but, unlike poststructuralist accounts, remains wedded to ontological realism. For a lucid review of various theories of knowledge as applied to understandings of nature, see Braun and Castree 1998, 3–42, "The Construction of Nature and the Nature of Construction."

12 Computed on the basis of rainfall figures for Matar town obtained in April 1995 from the Mahi Right Bank Canal (MRBC) Authority, Nadiad (Gujarat).

13 Cf. Clark 1979, chap. 8; Bates 1981; Hardiman 1981, chaps. 2 and 3. The term *Patidar* literally means one who has formal ownership rights or a *patta* (deed) over a piece of land; in Kheda the label came to signify wealthy landowners or estate holders.

14 Shah et al. 1990, p. 36, table 2.9; MRBC Authority, Nadiad, April 1995.

15 The distinction between "caste" and "tribe" was muddled; the terms were often interchangeably applied by colonial officials to groups like the Vaghris who economically, socially, and spatially inhabited the fringes of Gujarati rural society. For an interesting discussion on the caste/tribe divide, see Skaria 1997, 726–45.

16 Hardiman 1981, 47.

17 As John B. Thompson observes, ideologies of domination circulate in ordinary usage "as utterances, as expressions, as words which are spoken or inscribed. . . . to study ideology is, in some part and in some way, to study language in the social world. . . . the ways in which language is used in everyday social life, from the most mundane encounter between friends and family members to the most privileged forums of political debate" (1984, 2).

18 Actual names of individuals and villages have been altered.

19 Sabarbhai Tadbda, interview by the author, 2 May 1995, and field notes, 14 May 1995.

20 David Hardiman (1981, 44) makes a similar observation; also see Pocock 1972.

21 Figures obtained on 28 August 1995, Sub-District Registrar's Office, Matar.

22 According to Clark (1979) and den Tuinder (1992), it was common practice for village headmen *(patels)* to transfer usufruct rights over plots of land to themselves or to supporters under the pretext of raising cash to cover shortfalls in annual tax dues assessed on the village. Such lands were then classified as *gerania.* Colonial officials sought to eliminate this category of land because they considered it *khalsa* (government) land illegally alienated by unscrupulous headmen. The modern practice of *giro* is genealogically linked. Mario Rutten (1995), who did fieldwork in nearby Anand subdistrict of Kheda, documents its modern prevalence.

23 However, the weight of this informal rule diminishes if the borrower invokes the stricture against dispossession of ownership by requesting that the payment date be rolled back. The lender's response, of course, varies — in large part with the level of social opprobrium or disaffection he thinks foreclosure of his loan will invite. If the lender and borrower are from the identical caste group, the compulsion to be generous in negotiation (i.e., the fear of opprobrium) will be proportionately greater, although it deserves to be said that the level of opprobrium — or indeed, whether it is manifest at all — can vary with the borrower's economic condition, whether his inability to repay has justifiable cause, and whether the lender is thought to be justified in seeking foreclosure. It should be added that individuals who fashion themselves as leaders *(agevans)* and patrons *(daataas)* in the village are under greater compulsion to display generosity in transactions because the social legitimacy of their positions comes at a price, which in turn implies that public scrutiny of their actions — and, correspondingly, public derision at perceived transgressions of obligation — is sharper.

24 My intention here is to signal the theoretical significance of space in explanations of agrarian change, building on a more general framework of capitalist development proposed by David Harvey (1985). Other exemplars in this tradition are David Harvey (1989); Edward Soja (1989); Eric Sheppard and Trevor Barnes (1990); and Doreen Massey (1994).

25 See Harvey 1985 for a strenuous defense of this point. From a different political spectrum, literature rooted in the new institutional and information economics predicts that agents who face pervasive fluctuations in their real incomes will devise systems of insurance that include strategies of spatial diversification and transfers to smooth their income profiles, subject to constraints arising from moral hazard, adverse selection, transaction costs, and enforcement. See, for instance, Eswaran and Kotwal 1990; North 1990; Alderman and Paxson 1992.

26 See Marx [1844] 1988; Lukács [1968] 1986; Meszáros 1970; Habermas 1973, chap. 6; Aggers 1992; Postone 1996.

27 Demeritt 1998; Smith 1998.

28 The two most prominent advocates of "artifactual constructivism" are Bruno Latour (1987) and Donna Haraway (1991).

29 Bhaskar (1993, 228–30) deploys these terms in an explicit critique of the "Cartesian-Lockian representationalist view of knowledge," which, whether positivist, pragmatist, or superidealist, ultimately disengages the subject from practice and preserve an individualist-atomist conception of society. Bhaskar instead maintains that knowledge is socially *produced* and that reality, although "concept-laden," is not exhausted by those concepts (ontological realism); he also maintains that we negotiate reality *through mental or physical labor by referentially detaching it.*

30 Mann and Dickinson 1978; Goodman et al. 1987.

31 In Ted Benton's (1996, 172) eloquent formulation: "In any realist or materialist approach, enabling conditions must be understood as simply the obverse side of the coin from limits or constraints. A power conferred on human agents by a specific social relation to a natural condition or mechanism [i.e., nature's work] will also be bounded in its scope by that self-same relation."

32 The Kolis were subdivided into nonendogamous but stratified subgroups; these were respectively, in descending order of accepted lineage purity from Rajput ancestors, the Baraiya, Tadbda, Chuwalia, and Patanwadia Kolis. The subdivisions among Kanbis were more complex: the Lewa, Kadva, and Matia subgroups existed in a shallowly graded economic and ritual hierarchy that was partially endogamous. The Patidars, although racially affiliated with the Kanbis, occupied a distinctive economic and cultural stratum (akin to the English landed gentry) and at least in the early years of colonial rule appear to have been an endogamous subgroup. However, because members of the Koli community could also be Patidars, the issue of racial affiliation is a little more complicated than I have presented it. For details, consult Pocock 1972; Naik 1974; and Clark 1979.

33 Reginald Heber, bishop of Calcutta (1828). R. E. Enthoven (1922) relied primarily on Heber's account to describe Kolis.

34 Clark 1979, chaps. 1, 4, 8, 12, and 14. For instance, on pp. 83–95 she describes how Kolis and Kanbis were virtually indistinguishable to early colonial observers.

35 This is documented in chapter 4 of my doctoral dissertation (Gidwani 1996). Also see Clark 1979, chaps. 4, 7, and 8; and den Tuinder 1992, chap. 4.

36 Cf. Rabitoy 1975; Ballhatchet 1957.

37 Hardiman 1981, chap. 2.

38 Otherwise known as the *charotar.*

39 Hardiman 1981, chap. 2; also see chap. 4 in Gidwani 1996.

40 Hardiman 1981, 281, "Glossary"; field notes.

41 Harvey (1985, 153–57) terms this the "spatial fix." New Institutional Economics (NIE) and finance economists call it "exploiting gains from arbitrage," but make no explicit mention of the fact that arbitrage assumes the existence of a space-economy (as opposed to the reified space-*less* economy normally found in rational choice models).

42 Space limitations prevent me from pursuing this argument here, which is elaborated in my unpublished paper "Seeds of Indeterminacy: The Economic and Ecological Underpinnings of Agrarian Change in Kheda District, India."

43 The idea of "legibility" is borrowed from James C. Scott (1997). Hardiman (1981, 42) says more about the 1931 Census.

44 Literally, the "famine of '56"—so called because it occurred in the Hindu calendar year Samvat 1956.

45 Details of the famine and its aftermath in Matar are recounted in my unpublished article "Seeds of Indeterminacy."

46 While it is clearly advisable to read colonial archives with caution and circumspection there is no reason to believe that colonial officials unjustly lauded the agricultural prowess of Kanbis. Virtually all historians of central Gujarat concede that the Kanbis were in fact skilled cultivators; but they note, as do I, that the Kanbis' ability to practice a profitable agriculture owed greatly to their superior access to inputs, especially credit.

47 I discuss the ramifications of this in my article "The Quest for Distinction: A Reappraisal of the Rural Labor Process in Western India," *Economic Geography*, 76 (2): 145–68, 2000.

48 Kumarappa 1931; Desai 1948.

49 Ibid. for details.

50 K. D. Patel, conversation with the author, village Astha, 24 July 1995.

51 Change in paddy acreage computed from data obtained from Shah, Shah, and Iyengar 1990, p. 42, table 2.15; and Kheda Jilla Panchayat, *Jilla ni Ankdakiya Rooprekha*, 1989–1990, p. 41, table 4.5.

52 Kheda Jilla Panchayat, *Jilla ni Ankdakiya Rooprekha*, p. 77, table 5.

53 A trend of rising real wages for agricultural operations is illustrative. Thus, in Matar Taluka, the daily real wage for weeding, an operation that is common to all cereal crops, increased from 2.22 kilograms of paddy in 1965 to 2.39 in 1974, and 6.41 in 1994. There have been other shifts in labor contracting as well, which collectively point to an increase in labor's bargaining power. For instance, in peak agricultural periods, employers often have to recruit laborers for work the night before and by paying the daily wage in advance.

54 In the most detailed study of the impact of tractorization, Hans Binswanger (1980) concludes that there is no conclusive evidence for tractors having displaced laborers and worsened their bargaining power. Terry Byres (1981) has challenged this claim.

55 Cf. Bourdieu 1977 and 1990.

Reflections

David Ludden

Agrarian Histories and Grassroots Development in South Asia

This volume is about development (among other things), yet it arrives when prominent scholars are saying that development is dead.[1] Still very much alive, however, development institutions, ideologies, and disciplines continue to propagate modernity as they confront a vital transition. Social sciences have engaged development for more than a century, and by the 1920s, this engagement had formed academic disciplines concerned with improving the future of agrarian environments (Ludden 1994, 1–35). By then, the idea was well established that "development" referred to officially sponsored and patronized (if not organized) efforts to improve living conditions, especially among the poor, going beyond charity to include overall moral and material advancement, combining elements of public works, infrastructure building, famine prevention, disaster relief, social reform, and welfare legislation (Ludden 1992). William Moreland, a Mughal historian working in India in the 1920s as an agricultural officer, argued that development imperatives had been part of Indian statecraft from medieval times (Moreland 1929). Like Moreland, many critics have long attacked governments for failing development (Ludden 1999, 6–17, 180–217), and critics of the British empire such as Jawaharlal Nehru became national leaders in the global development regime that emerged from decolonization after World War II. This regime crumbled, bit by bit, between 1979 and 1991. Its Cold War context—composed of oppositional state powers promoting capitalism and communism—fell apart. Its major capitalist institutions (the World Bank, IMF, foundations, multinational corporations, and expansive national economies) defeated their opposition and announced a new dawn of globalization. At the same time, the state lost legitimacy as a development institution. Popular participation gained new authority, along with institutions such as markets and nongovernment organizations (NGOs), which conceal power and inequality in the flurry of exchange, pricing, voluntarism, and local accountability. In academic literature, venerable assumptions about who would develop what, for whom, and to what end came under attack, along with the idea of modernity itself. States came to appear as inherently rigid, stupid, and oppressive. As Ronald Reagan became president and Reza Shah Pahlavi fled Iran, the new Right fought to "get the state off our backs," and the new Left fought against state control of development. When communist states collapsed, state-led revolution became an oxymoron. In this climate, "development" became identified with projects to use economic growth against revolution, as Robert McNamara had done at the World Bank. Managerial state mentalities lost authority in development thinking, and free markets, worlds without borders, and insurgent localism rose to prominence. Subaltern representation, choice, and empowerment redefined the who, what, and wherefore of development, and efforts to improve everyday life "from the bottom up" preoccupied new research. This transition frames the essays in this volume.

Two types of power shifts are changing the meanings and forms of development today: they are political and intellectual. The new social movements have abandoned revolution as an ideal for oppositional activism seeking structural transformation, and national states are losing control over national economies. As a result, thinking about development is more detached from the state than ever before and moving in two opposite directions: outward into networks of globalization and downward into localities of grassroots activism. The chapters of this volume follow the latter trajectory and reflect a related shift in academic practice. The social movements concerning gender and the environment that most influence the authors here have produced a huge political space where Left and Right polarities lose their potency. Human rights, gender equality, and ecological sustainability dissolve Left and Right loyalties to instigate promiscuous recombinations of their various elements. Likewise, globalization leads scholars across political and academic boundaries (Appadurai

1996). Grassroots commitments are also reshaping knowledge for new constituencies and agendas that do not obey the old geography of ideas. In this book, agrarian environments are constructed conceptually of material drawn from ecology, political economy, anthropology, history, economics, and cultural studies. The primacy of real-life problems, and of local, everyday experience, outweighs the authority of academic disciplines (Andranovich and Lovrich 1996). Authors argue that new kinds of knowledge are needed to address development problems in our contemporary context. Perhaps the implication is that political economy has been shaped intellectually by its political utility for competing ruling elites. Certainly, agrarian studies that center on states, markets, productivity, policy, and class power have failed to grasp basic grassroots issues, notably, those concerning gender and ecology.

Local activists now deploy a range of ideologies that do not fit into the Cold War template of the old political economy.[2] NGOs play the most prominent role in grassroots development, and many take a positive attitude to world markets and to capitalist enterprise. Most are not remotely revolutionary and work to change the world incrementally by making a real difference in local settings and delivering benefits to local constituencies. Grassroots issues and nonrevolutionary but progressive human agency are now major academic concerns, and scholars who might have once looked for the roots of revolution in contradictions of capitalism now study localized, often doggedly individualistic struggles among permanently subaltern folk (Scott 1985; Haynes and Prakash 1992). This volume emerges from ongoing efforts by social scientists to improve their disciplines by keeping their knowledge up-to-date and useful, and to this end, they need to change its form and content. Here, the term "environment" encodes the intellectual heritage of environmentalism, as "agrarian" does the legacy of agrarian studies, to form a composite idea of agrarian environments, combinations of localized power, production, culture, and nature. These authors participate in the reconstruction of development discourse after years of incessant critique, and the volume also strives to provide appropriate academic knowledge for development activism.

Its combination of empiricism and critical theory indicates a new style of development discourse, whose goal is not to instruct policy elites but to instigate popular participation and to insist on the inclusion of marginal peoples. This move to the margin is part of the political and intellectual shift away from centers of state power. What makes development different from economic growth is that development includes a public discussion of goals, means, and values, a collective reflection on past experience, and a consideration of alternatives. Development discourse is public debate and shared knowledge. But who participates? Who talks and listens? Who is developing and who is becoming developed? These are now central questions. Social movements and their allied scholars assert the right of poor and marginal people to participate, to produce development knowledge, and to control development, rather than merely to fight or criticize the power of the state. Broadly inclusive popular participation is now the accepted norm. Unaccountable planning and elitist control—by corporations, state officials, and technical experts—is no longer acceptable. Democratic development is widely promoted, much more by scholars and activists than by governments and funding agencies; this volume thus has a real political location and meaning. This research argues that local experience, local knowledge, and local participants should play a more prominent role in development than grand theories, state interests, and world markets. How to make this happen is not clear. But clearly scholars can contribute by reporting, analyzing, and debating development, by making public the information that people need to participate.

As in the new social movements, in this volume, the state remains a lurking presence. Whereas twenty years ago authors would have focused on questions of state policy, now the concern is to interrogate the local features of state power.[3] This reflects the shift in academic practice noted earlier. Historians have largely abandoned the state and moved into social and cultural studies, and anthropology and sociology have become more concerned with local material conditions, and political scientists do more fieldwork (e.g., Fernandes 1997). A new perspective on the state has emerged in which we can see state power and social forces constituting one another (Migdal et al. 1994). At the same time, the autonomy of national states continues to decline. In India, the national political arena has fractured into kaleidoscopic alliances (Kohli 1990). New social movements and Hindu majoritarianism have splintered national identity (Bonner 1990; Omvedt 1993; Ludden 1996c), and structural adjustment and liberalization prevent the state from making firm development commitments. Regional, popular, and global forces bounce development in various directions. Development projects are now most often outside state control. Yet national states also authorize most development projects; national boundaries inscribe the public sphere; and national systems of law, politics, and culture implicate every locality. There is no escaping the state. But its role and its place in scholarship are changing, and several chapters in this volume indicate how social scientists are putting the state under a new kind of scrutiny.

As national states lose control over development, and as projects disperse into global localities, development discourse loses the coherence that it has when trapped in national policy debates. It is much easier to organize a development debate around nationalist or revolutionary slogans than around the subtle complexities of feminism or environmentalism. When grassroots issues take precedence, it becomes harder still to compile data for a broad range of local cases to formulate a general picture of what is going on. Volumes like this one that seek to mediate between sophisticated academic research and everyday development debates need to organize their information accessibly; and in addition to focusing on agrarian environments, gathering a set of case studies from India makes good sense, because national territory remains the basic unit of global aggregation. Web sites for UNICEF (http://www.unicef.org/statis/), UNDP (http://www.undp.org/hdro/indicators.html), and many other agencies indicate that national territories are still the most widely used for global measurements and comparisons, however strange it might be to compare the Human Development Index for countries such as India and China that have hundreds of millions of people each with the sixty or so countries that each have fewer than five million inhabitants. Such empirical use of national territories presents a conceptual problem because it implies a central position for a national state in defining the character of localities, but clearly that is not intended in this volume. But what, then, is the status of "India" here? In the individual chapters, the Indian *state* does appear now and then as a feature of local power. But it does not appear as an integrated institutional influence on these cases, holding them all together.

For the remainder of this short ancillary essay, I want to argue that these case studies do inhabit a distinctive agrarian region of the world, which forms a useful context for their interpretation. For all these environments, "India" does not only mean the territory that happens to be enclosed by the Republic of India, or merely connote the Indian state; it also denotes a dense spatial combination of commonalities and connections that are of practical importance in everyday agrarian life. Basic commonalities include environmental elements like seasons and monsoons. In more northerly regions of the world, temperature changes between winter and summer determine agricultural seasons, but in India, Nepal, Bangladesh, Pakistan, and Sri Lanka, monsoons define seasonality. "India" is a distinctive kind of natural environment, which lies south of a markedly more arid environment in western and central Asia, and east of the humid tropics in southeast Asia. Indian agrarian environments—a phrase that covers the Indian subcontinent, or south Asia—display seasons, flora, fauna, topography, and other natural elements that differ visibly from neighboring regions in Afghanistan and Burma, Tibet, and Indonesia.

Historical patterns and forces of change also connect south Asia's agrarian environments more densely to one another than to places across the Indian Ocean, across the plateaus of Iran and central Asia, or across the mountains between Assam and China. All the places in this book participate in the histories not only of the Republic of India but also of the British and Mughal empires. The sixteenth century is a good place to begin tracing influential commonalities formed by imperial institutions. Mughal accounts of revenue assessment, collection, and entitlement indicate that by 1595, many regions in south Asia were connected to one another as objects of managerial intelligence. Abu-l Fazl was an early state intellectual concerned with development. His *Ai'n-i Akbari* composed a landscape of revenue statistics, and his *Akbarnama* provided a charter for Akbar's authority in agrarian territories (Habib 1982, 1995; Moosvi 1987; Chetan Singh 1995). Other texts give a glimpse of how contested Mughal managerial intelligence was in practice. The *Tarikh-i Hafiz Rahmat Khani,* for example, explains that Pakhtun tribes need to be pitted against one another to ensure that "Pakhtun will hit Pakhtun" rather than fighting the Mughal sultan (Nichols 1997). At every level of political authority—inside and outside Mughal territory, from peasant household to sultanic durbar—ranked individuals measured their own agrarian environments with revenue payments up the ladder of imperial ranking. Genealogies and hagiographies defined entitlements in transactional ranks that constituted agrarian environments in early modern political culture.[4] Abu-l Fazl constructed an imperial environment by accounting for all its economic resources. Cavalry, guns, robes of honor (S. Gordon 1996), statistics, patriarchy, and commercial finance defined powers in agrarian environments for people who controlled local resources, both physically and culturally. It can be said that today's intellectual interventions in development debates are but the most recent efforts to redefine local authority in Indian agrarian environments. Struggles to control nature have a long history, and there is some specificity to the terms of these struggles inside the territories of Indian states.

Struggles over land revenue and entitlement dominated the politics of agrarian environments until about 1850. Like Abu-l Fazl, the East India Company also thought about agrarian improvement in terms of revenue, rights, and titles to land.[5] Company Raj did produce a new kind of agrarian ethnography. It did make its own entitlements supreme. It did analyze economic resources, technologies, trade, and occupations more deeply than any previous state.[6] But like its predecessors, the Company paid very little attention to the practical details of resource use.[7] Literary genres (from Tamil *pallu* and *kuruvanji* to Pakhtu *ghazals* adapted from Persian) attest to astute agrarian observation, but the official study of agrarian environments did not begin until after 1850; then, modern bureaucracies for collecting revenue, adjudicating rights, counting people, describing groups, and measuring assets brought agrarian environments into cultural existence for modernity (Ludden 1985, 1996a; Nichols 1997, chap. 2). By the 1870s, when W. W. Hunter compiled his statistical accounts of Bengal and Assam, development had begun to preoccupy state intellectuals and middle-class authors such as Dadabhai Naoroji and Bankim Chandra Chattopadhyay. Novels, plays, and poetry began to record local life in detail. By the 1920s, the village had become a standard object of economic analysis, and in 1916 the academic field of Indian economics (described in its first textbook by Radhakamal Mukerjee) rested firmly on theories of traditional, local institutions, starting with religion, caste, and village society (Ludden 1994). By the twenties, city intellectuals had designed rural India for the modern mind with a classically ecological character: the Indian village formed a self-sustaining, homeostatic niche for the reproduction of interdependent popular folkways, inscribed in caste (*jati* and *varna*). The institutions of modernity—states, bureaucracies, newspapers, universities, and academic disciplines—were understood to be external and foreign to the natural ecology of Indian agrarian environments, but in fact, moderns invented traditional agrarian environments endowed with all the elements of culture, nature, technology, and religion.

This invention quickly attained a distinctively Indian form. Nationalized and embellished, it served to essentialize all the countries that emerged from and bordered on British India. Nationalist critics blamed the British for the poverty of rustic folk and for the disruption of village society. Imperialists blamed rural poverty and stagnation on pre-British regimes, wasteful natives, caste tradition, the weather, and many other factors. Such oppositions became part of the cultural construction of agrarian environments in south Asia and influenced all later debates about development. The village became the heartland of tradition, stability, and, by the 1980s, ecological harmony. Agrarian environments continue to symbolize the national essence in all the countries of south Asia, so that state development activity has taken on a contradictory double meaning. On the one hand, Nehru's kind of modernity machine became the new Indian nation producing development; on the other, it also became another modern state's assault on ecologies of local tradition. Grassroots environmentalism has gained passion and certainty from a cultural setting in which people know that marginal, rustic folk have always been oppressed by capitalists, imperialists, and foreigners; that traditional societies are ecologically sound; and that environmental ruin has taken the name of progress since the days of British rule.[8] National histories form such cultural settings.

By the 1920s, moreover, national movements were also dividing the land among conflicting indigenous representatives. Movements led by M. K. Gandhi, M. A. Jinnah, Fazlul Haq, Khalistanis, Jharkhandis, Telengana rebels, and many others now have given the nation many local and regional meanings that express a form of political and ethnic territoriality shared by most agrarian environments in south Asia. The land-in-the-blood thrives in national institutions, even as they often conflict. By 1931, when Jawaharlal Nehru became the president of the All-India Congress Committee, a national development regime had begun to overspread all local claims to land, asserting forcefully that the national state would supersede the dominant powers in all the many regions of ethnic attachment. His 1931 proclamation suggests how agrarian environments became the object of contesting normative claims:

> The great poverty and misery of the Indian People are due, not only to foreign exploitation in India but also to the economic structure of society, which the alien rulers support so that their exploitation may continue. In order therefore to remove this poverty and misery and to ameliorate the condition of the masses, it is essential to make

revolutionary changes in the present economic and social structure of society and to remove the gross inequalities. (Zaidi 1985, 54)

Many competing nationalist impulses—secessionist, revolutionary, religious, capitalist, ethnic, tribal, and neofeudal—have appeared in most localities, along with development institutions over which they stake competing claims. National struggles against foreigners reincarnate endlessly in battles led by grassroots activists. New social movements may appear to be (and often claim to be) manifestations of indigenous communities, but in this very imagery they indicate that none exists outside the institutional framework of the modern state. Conflicts over local control of environmental resources suffuse development and constitute environmentalism on the ground, as indicated by several chapters in this volume.

In their ecological histories, too, the agrarian environments in this book are distinctively south Asian (Ludden 1999, 48–59). High mountain regions along India's northern rim preoccupy Haripriya Rangan, Mark Baker, Vasant Saberwal, and Shubhra Gururani. Vinay Gidwani and Jenny Springer study irrigated agriculture along the coastal plains, in the west (Kheda district, Gujarat), and in the far south (Tirunelveli district, Tamil Nadu). Sumit Guha and Cecile Jackson and Molly Chattopadhyay examine landscapes of intersecting hills, valleys, and plains in the mountainous interior, where steady upland farmer colonization has displaced tribal and pastoral peoples—in the west, in semiarid Maharashtra; and in the humid tropics of the eastern Ganga basin, in south Bihar. Paul Robbins and Darren Zook work in what we could call India's badlands: Paul follows pastoral people who make their livelihoods in the arid desert fringe, in Rajasthan, and Darren follows famines in the dry peninsula.

All the authors discuss ecological change with extremely long-term parameters. Some common patterns of timing and causation influence all these agrarian environments. Technological change, especially in irrigation, urbanization, and industrial pollution, have had their most profound impact since independence in 1947 (and 1948 in Sri Lanka). Between 1901 and 1951, the work force became more agricultural in south Asia (when cultivators and laborers in undivided India increased from 69 to 73 percent of the male work force) (Krishnamurthy 1983, 535); but then the trend moved in the opposite direction, and today farming accounts for only 57 percent of the total work force in Bangladesh, 63 percent in India, 50 percent in Pakistan, and 43 percent in Sri Lanka.[9] During three decades after 1950, livestock, net cultivation, and built-up land increased as much as they had during seven previous decades, and forest cover declined at about the same rate, and population grew about 15 percent faster.[10] The influence of urbanization today goes well beyond the sprawling impact of huge cities, because regionally, urban growth is inverse to urban population: where urban centers were least prominent in 1901, their local expansion then became most far-reaching. The percentage of India's population living in urban centers increased by just over 1 percent (from 11 to 12 percent) during the first three decades of the twentieth century, by 6 percent during the next three (1931–1961), and by 8 percent from 1961 to 1991. This trend appears in all countries except Sri Lanka, which started with a relatively big urban population (12 percent in 1901) and now has less than twice that proportion (22 percent in 1991), whereas India's 1991 figure (26 percent) is 2.4 times what it was in 1901 (11 percent). Recent acceleration is quicker in Pakistan, where the urban population increased 70 percent faster than India's after 1961 to reach 33 percent of the total population in 1991. Nepal's small urban population (9 percent in 1991) has grown as fast as Pakistan's since 1961 (Schwartzberg 1992, 114, 280). Bangladesh is the most dramatic case. In 1961 its population was only 5 percent urban, which was only double the 1901 figure, but grew 400 percent after 1961 to reach 20 percent in 1991; and in the early 1980s, the urban growth rate hit 10 percent per year. Some of this increase resulted from reclassification. The 1981 Bangladesh census "extended the definition of urban areas to include small administrative townships and economically significant production and marketing centres . . . which had certain significant 'urban' characteristics." The number of urban centers increased from 78 (in 1961) to 522 (in 1991), and today more than 500 urban centers have populations of less than 5,000, while the four largest cities contain almost half the urban population, nearly 7 million people (Adnan 1998, 1338, 1347 n. 4).

Despite recent acceleration, environmental change has been synonymous with agrarian history, forever. The myth of ecological stability in premodern times is in startling opposition to what we know. A unity attaches to the history of all agrarian localities from the simple fact that expanding cultivation has been the primary means to increase farm output for as long as we know (Blyn 1966, 127). More intense cultivation to produce more crops on land already farmed means more frequent planting, more nutrient-demanding crops, more labor, and almost always irrigation; and such intensification affects the environment in complicated ways. Agriculture is by definition a forceful ecological intervention, and it always involves some kind of social conflict over control of the environment and fruit of the soil. Since Mughal times, forest cover and unfarmed land for grazing and other uses have decreased everywhere, and state power has been exerted very substantially to increase cultivation, to increase total agrarian wealth.

Recently scholars have paid more attention to this process, which enables us to form a comparative, long-term perspective on the agrarian regions that make up south Asia (Agnihotri 1996, 59–73). Using *A'in-i-Akbari* statistics, Shireen Moosvi (1987, 39–73) estimates that gross cropped area in the Mughal heartland covered 61 percent of the land that would be cultivated in 1910, but ratios varied from an average of 85 percent in Agra, Bet Jalandhar, Baroda, and Surat to 29 percent in Champaner and Rohilkhand and 8 percent in Sindh Sagar. James R. Hagen (1988) argues that in the Gangetic plain and Bengal, roughly 30 percent of the total area was occupied by farms in 1600, 50 percent in 1700, 50 percent in 1800, 65 percent in 1910, and 70 percent in 1980. By his estimate, cultivated acreage more than doubled between 1600 and 1910, with more than half the increase before 1700, none in the eighteenth century, and negligible amounts following 1910. In Punjab, the picture is quite different. Outside Bet Jalandhar, Punjab lowlands were barely cultivated in the sixteenth century; in 1800, most land south of the submontane tract was still open for grazing (Chakravarty-Kaul 1996, 35–63; Agnihotra 1996), but between 1850 and 1939, the government built 20,886 miles of canals, which by 1945 irrigated 15,688,000 acres, much of it bearing more than one crop.[11]

Regional disparities of this sort typify south Asia. Outside Mughal territory, higher proportions than 50 percent of 1910 gross cultivation would have pertained in old areas of intense cultivation along the coast and in the Ganga basin. Moosvi's figures for Baroda and Surat probably reflect conditions on the coastal plains, especially along riverbeds and in the deltas (except in Bengal). As Hagen's estimate suggests, the increase would have been small in the eighteenth century, during wars, plagues, and famines, particularly in the later decades. Bengal took fifty years to reclaim land lost to famine in the 1770s. In Madras Presidency, expansion from 1800 to 1850 substantially involved bringing old farmland back under the plow. In 1850 vast areas for new cultivation remained in Bangladesh and Assam, where farming frontiers continued to move east, south, and into the hills; and also in Punjab, Haryana, Gujarat, and western Ganga basin, where large-scale irrigation was installed, as it was in the deltas of the Kaveri, Krishna, and Godavari Rivers in the southeast peninsula. In old agricultural territories, however, there was little expansion after 1850, though as Cecile Jackson and Sumit Guha indicate, even incremental change would displace hill peoples and forest ecology; this occurred throughout the nineteenth and twentieth centuries in India's central mountains, high mountains, and Western Ghats.

We can sketch the long-term interaction among productive activities, ways of life, and ecology roughly as follows. Something more than half the farmland that produced crops in 1900 was being farmed in 1600, and the oldest of this farmland was more than a thousand years old, and almost all of the oldest farmland was clustered in riverine lowlands and deltas. Islands of farming formed a medieval archipelago dominated by forest and scrub and by pastoral peoples, hunters, and many scattered farm communities. Farm expansion accelerated dramatically after 1600, when the second half of the land area cultivated in 1910 was carved into the earth in just 300 years. Dry regions of the inland interior were still on the whole very sparsely populated in 1800, however, and they still contained substantial pastoral populations. As Sumit Guha shows, the nineteenth century witnessed a rapid increase in dry cultivation and resource scarcity. The increasing frequency of agrarian unrest after 1850 hinges on competition over land, water, farm income, and rights to resources amid the final closure of farming frontiers. It also brought large irrigation works to water vast dry plains. Thus, amid an acceleration of trends under way since 1600, regional disparities grew. As irrigation made Punjab rich, the desiccation of the peninsula interior produced landscapes of drought, hunger, and famine.

Since 1880, ecological change and human dislocation has concentrated in the high mountains and western plains, including Rajasthan and parts of Gujarat. By contrast, Jenny Springer's Tirunleveli district has seen relatively little disruption. In India as a whole, the ratio of total farmland to total land area in 1990 as a percentage of this same proportion in 1880 indicates high rates of increase (and thus rapid decline in forest cover) in Tripura (903 percent), Sikkim (698 percent), Nagaland (405 percent), Assam (333 percent),

Rajasthan (326 percent), Mizoram (288 percent), Arunachal Pradesh (271 percent), and Orissa (206 percent), compared to low figures running from 103 to 122 percent in Tamil Nadu, West Bengal, Uttar Pradesh, Maharashtra, and Kerala.[12] Expansion (with new irrigation) in Rajasthan, Haryana, Punjab, Gujarat, Karnataka, and Sind, and (with forest clearance) in mountainous tracts in Orissa and Madhya Pradesh, is substantial. The most dramatic change, however, is on the farthest frontier of long-term lowland expansion, in the high mountains and Assam. The colonization of these areas from the lowlands did begin under the Mughals, and the exploitation of their forest resources did increase after 1800. But rapid acceleration set in after 1880 and continued after 1947. The percentage of land under cultivation at high altitudes remains low, so agricultural expansion still has a long way to go. Rapid proportionate increase of farm acreage, and decline and degradation of forests, helps to explain the steady increase today in conflict over land and resources at high altitudes.

It is no wonder, then, that mountain ecologies are the most pressing concern for environmentalists. Perhaps the most impressive contribution of this volume is to extend this concern into the lowlands, from the historical margins of agrarian history into its heartland. The next step is to extend this model of research into zones of urbanization and the industrialization of agrarian environments, where the great bulk of ecological disruption and the most severe of the "gross inequalities" of power bear down on the countryside.

NOTES

1 See Escobar 1992, 1995; and Sachs 1992, which opens with this: "The last 40 years can be called the age of development. This epoch is coming to an end. The time is ripe to write its obituary."

2 For a sophisticated effort to renovate agrarian political economy, see Harriss-White 1996.

3 I am indebted to one of the anonymous readers for Duke University Press for this wording.

4 For background, see Thapar 1978, 240–361; and Chattopadhyaya 1994. For texts, see Ferishtah 1792 and Elliott 1966.

5 See *Risala-Zira't* (Treatise on agriculture) discussed and translated in Mukhia 1993, 259–94.

6 For a consideration of some of these data, see Ludden 1996b.

7 As a result, historical reconstructions of agrarian environments before 1850 need to rely heavily on later data. See Chakravarty-Kaul 1996, 35–62; and Ludden 1985.

8 It is logical, then, that to launch the study of Indian environmental history, Ramachandra

Guha and Madhav Gadgil (1992) recast India's colonial impoverishment as a tale of ecological devastation.

9 It is 93 percent in Nepal. See B. Agarwal 1994, 51, which uses World Bank statistics.

10 For data on thirteen countries in south and southeast Asia, 1880–1980, see Richards and Flint n.d. Data from this study are available on the Internet.

11 *Administrative Report of the Punjab Public Works Department* (Irrigation Branch), 1945–1946, in Islam 1996, table 1.1.

12 This paragraph is based on calculations from official data compiled by John F. Richards and his colleagues for the period from 1880 to 1980. Data from this study are available on the Internet.

Cathecting the Natural

There is a pervasive way of thinking about the relationship between modernity and the natural or nature. By its lights, premodern societies had a balanced and relatively nonexploitative relationship with nature. In contrast, in modern societies, a more mechanical relationship with nature and the natural developed. A split was instituted between the natural and the cultural, and the former was feminized, exteriorized, and subordinated. This way of thinking, which can provide a powerful critique of modernity, has framed much environmental history. In the South Asian context, it informs arguments that women or precolonial or rural societies have a more harmonious relationship with the natural—consider the shared element in the otherwise very different work of Madhav Gadgil, Ramachandra Guha, and Vandana Shiva.

Increasingly, however, these once-radical arguments seem untenable. Their claims to precolonial, premodern, or feminine harmony have been undermined, and rarely as effectively as in this volume—I think especially of the papers by Mark Baker, Sumit Guha, Shubhra Gururani, Cecile Jackson, and Molly Chattopadhyah. So have the claims that colonial or postcolonial modernity necessarily leads to ecological degradation or environmental violence: it is precisely the concern to undermine such teleologies that inform Gidwani's remarks about technological determinism, or, in a different way, Paul Robbins's arguments about the varieties of pastoralism.

All of this raises the question: how do we think of colonial or postcolonial modernity and their relationship with the environment? One answer, frequently resorted to, is to insist on the recognition of multiple interests, heterogeneity, and fuzzy boundaries. Although this is of course important, such a recognition sometimes runs the risk of becoming an anodyne liberal empiricism or pluralism where all perspectives are equal because partial. I would therefore like to take my cue from the papers in this volume (and especially from Saberwal's emphasis on the inescapability of constructions of the natural) and chart, in a hesitant and tentative way, the cathexis of nature and the natural in colonial and postcolonial practices. (I shall refrain from an enumeration, analytical separating out, or specification of the many meanings of ideas of nature and the natural, for the power of nature and the natural as signifiers may in large part be due to the slipperiness and ambiguity of their meanings in everyday practices.) In talking of cathexis, I am not simply making the point that the natural is imagined—like communities, nations, or traditions are imagined. I talk of cathexis because I want to foreground the particularly charged importance of these imaginings. The imagining of nature was not just one more set of social practices. It was rather a crucial dimension of the anxiety and uncertainty involved in the constitution of the modern. By focusing on the cathexis of nature, I would like to open up for consideration the possibility that the problem with scholars like Guha or Shiva is not that they go too far, thus obliterating with their extremism some heterogeneous liberal median; the problem is rather that they do not go far enough and thus remain implicated in a peculiar modernist cathexis of the natural.

ONE

Maybe a point of departure for considering these cathexes can be the relationship, for colonial officials, between the colony and the natural. Zook has argued in his paper that famine and poverty were staged as manifestations of the natural, of nature, and as a problem of the landscape. Once staged in this way, as Zook remarks, the task of famine prevention was to protect "particular tracts" from the irregularities of nature, and to bypass nature with the use of science and engineering. Zook's point can fruitfully be extended elsewhere. In Europe, too, since around the seventeenth century, new representations of poverty and famine had been developing, ones that saw them as manifestations of the natural but at the same time sought to transform them. What was distinctive about the imagining of the natural in the colonies, then, was that here it was not only famine and poverty but the colony itself that was staged as a manifestation of the natural.

Thus there was what Partha Chatterjee (1993) has called the rule of difference. Colonial rule, he suggests, was premised on the positing by the British of an ineradicable distance between colonizers and colonized. This distance legitimated colonial rule, justified it both as domination over those incapable of ruling over themselves and as a civilizing mission that sought to at least reduce the distance. In our context, what is relevant is that this rule of difference was articulated and enacted not only through imposing various laws and practices but by identifying the colony and the colonized with the natural, and the British with that which was outside the natural and yet intimately linked to it.

Consider how the trope of the natural was involved in British understandings of communities that came to be called the castes and tribes of India (for a detailed analysis, see Skaria 1997, 1999). The tribes were regarded in colonial discourse as masculine, and as in this sense having affinities with imperial masculinity. But they were separated from imperial masculinity by evolutionary and psychological time and were identified with natural man. Indeed, they were often explicitly described as living in a state of nature or as being childlike men. That is to say, the natural staged by the masculine tribes was a time that had been passed through by the colonizers; this natural was the past of the colonizers.

It is important to remember that new ways of thinking of the natural, and natural man, had emerged in the eighteenth and nineteenth centuries. As Anthony Pagden (1986, 8) has pointed out, formerly natural man was not so much a child of nature as "someone who had chosen to live outside human community." The trope of the tribes in colonial thought, one might say, involved a cathexis of the natural as the masculine prehistory of modernity. In contrast, the castes' relationship with the natural was constructed not through the medium of time but through associations of the natural and the feminine. In the dominant conventions of Western thought, modernity has been viewed as distinctly masculine both as space and time; femininity itself was often viewed as natural and outside modernity (see, for example, Felski 1995). In this context, it is surely significant that the castes were repeatedly portrayed as effeminate, that tribe was often to caste as male was to female in colonial understandings (Skaria 1997). This femininity of the castes marked their proximity to the natural.

What makes the imagining of the colonized as natural especially charged is that this move was crucial in the constitution of the modern: it was by defining the colonized as natural and thus as the other of the modern that the modernity of the colonizers was made visible and claimed. Furthermore, this distancing from the natural was a deeply anxious one, always marked by the fear that the distance between the natural and the modern would be erased in ways that called into question the modernity of the colonizers. Hence the apprehension, well explored in academic literature, that contact with the colonizers would make the colonizers degenerate—usually effeminate or savage; hence also the British efforts to cast revolts by tribes as caused by boyish boisterousness or by the manipulations of devious caste outsiders, for the affection of the tribes affirmed British masculinity, and to conceive of genuine hostility from the tribes would have threatened to render incoherent the British claim to masculinity.

TWO

Colonial officials not only cathected the colonized as natural; they also sought to transform this natural through what David Ludden has felicitously described as development regimes. Among the aspects of development regimes that Ludden notes, three are of special interest here: a "people" who must be improved, a state that is regarded as the agency for effecting this improvement, and the claim by the state that progress is its goal and reason for existence. Progress, of course, is about modernity's particular kind of relationship with time: to be modern is to claim to be at the cutting edge of time, to be contemporaneous with evolutionary time. Development regimes effectively involve efforts to bring about progress by transforming this natural and the time it occupies—efforts, in other words, to make the natural modern. The activities of the development regime that Ludden describes—surveying, mapping, triangulating, conducting censuses of people, livestock, and agricultural instruments—were about the staging of landscape and people as a natural that had to be transformed, reworked, and bypassed to bring about progress.

It is sobering to recall how the very subject of this volume—the agrarian environments of south Asia—has been shaped, maybe even created, by the British development regime. In eighteenth- and early-nineteenth-century India, as much recent research has shown, settled agriculture was quite inseparable from other activities such as pastoralism, shifting cultivation, hunting, and raiding; even small forests were quite common in large parts of plains India. In such a context, agrarian environments were at best spaces that were constantly traversed by other practices. But Pax Brittanica, in the name of progress, changed all this: it promoted settled agriculture and actively suppressed shifting cultivation, pastoralism, gathering, and other related modes of livelihood.

As a concept, the development regime is also helpful in thinking of post-colonial cathexes of the natural. Jenny Springer's paper has reminded us of how the Indian state continues to be posited as the motor for rational action and progress, and how this positing is crucial to the constitution of both state and society. Through the practices of the state, farmers and local agricultural practices continue to be staged as the natural, as objects to be transformed by the practices of the development regime. Of course, there are profound departures from the colonial regime: the rule of difference is no longer constituted by the ineradicable distance between colonizers and the colonized, but by the shifting goalposts of what Dipesh Chakrabarty (1992) has so evocatively called the hyperreal Europe. The hyperreal Europe is a reified figure of the imagination that hypostatizes an idealized European experience into a universal set of criteria. By the criteria this hyperreal Europe generates, both state and society are defined by a lack and inadequacy, in constant need of reform and progress; in this sense, as Springer notes, the Indian postcolonial development regime is about transforming not only "people" and landscapes but itself. Put another way, given the lack made visible by the hyperreal Europe, there is a sense in which the development regime itself is natural. Thus there is no clear locus in the state for a development regime; the development regime is intimately linked to the state but is not the same as the older notion of the state. This older notion involves understanding relations of power in juridico-discursive terms, in terms of an opposition between state and society; the productive power of the notion of a development regime lies precisely in the fact that it enables us to sidestep these terms.

THREE

From what I have said so far, it might seem that for development regimes, the natural can exist only as a moment of resistance to be overcome, as a resource to be transformed, as an other to be subordinated. The persuasiveness of such arguments may derive from their convergence with a well-established trope in social theory, associated most of all with Hegel and Weber, that sees the growth of modern society as the growth of rationality, the state or bureaucracy as a particularly important locus of that rationality, and the natural as the object that this rationality acts on. In recent times, these arguments have also been associated with a dissatisfyingly straightforward reading of Foucault—should one say an elision of Foucault with a particular strand of Weber?—where development is regarded as erasing politics while simultaneously pursuing the very political task of expanding bureaucratic state power.

These arguments are not wrong. But perhaps they need to be supplemented with the recognition that development regimes, or the practices of development, also involve a certain primitivism. This primitivism is an attempt to gloss over a constitutive aporetic moment in the extension of rational or bureaucratic power. The primitive, recall, is also the natural; in this sense, as well as in its importance and in the anxiety surrounding it, primitivism is inseparable from development regimes, and from modernity itself. Involving as it does a powerful range of practices that produce, sustain, and affirm the natural as a past made visible in the present, primitivism renders contemporaneous the primitive and the civilized; it is an affirmation by the civilized of the primitive.

Colonial primitivism was of course most spectacularly visible in the British celebration of the wild masculinity of tribes, in their emphasis on preserving the noble savage from the effeminate castes. But primitivism was also involved in British Orientalism, if in less dramatic ways. True, Orientalist scholarship claimed to be classicist (in the sense that the classic is that which claims to be unmarked by, or indifferent to, evolutionary time; it is that which is located in an eternal present) rather than primitivist. Yet as Said's argument suggests, Orientalism could never be classicist: it was inevitably about locating the Orient in a time different from that of the West, a time behind that of the West. Furthermore, this Orient had to be produced in the present, and it is in this sense that I talk of colonial primitivism.

One of the most fascinating dimensions of the chapters in this volume is their focus on the technologies of primitivism, such as the extension work analyzed by Springer, or the *Riwaj-e-Abpashi* analyzed by Baker. Consider the *Riwaj*—the record of rights to irrigation from old water channels that the British created. It is conventional, of course, to read colonial writing such as this as informed by exactly the opposite impulse of primitivism; we normally see it as part of the development regimes of rationality, as introducing fixity where there was fluidity, as rendering, shall we say, illegible nonstate spaces legible. This is quite correct, but is it adequate? The *Riwaj-e-Abpashi,* after all, did not simply reduce rights and create order; it cannot be seen simply as a move from fluidity and complexity to fixity: disputes continued, and the *Riwaj* itself had to be constantly supplemented by judgments. Nor, of course, is it adequate to talk simply of a shift from fluidity to fluidity, or complexity to complexity—as though nothing had changed at all. Perhaps the transformation betokened by the *Riwaj* can be understood along two registers. One could focus on the emergence of writing as the medium through which complexity and fluidity were now created and contested, and the ways in which this process marginalized subaltern groups. The second is more directly related to primitivism as modernity's product, and as a gloss over its aporias. The *Riwaj,* in codifying rights, also effectively gestured toward their otherness, the fact that they resided in custom, not law; in practice, not rationality. The *Riwaj* was intended not so much as a denial of what officials saw as custom as a reification of custom, an acknowledgment of its prior nature and irreducibility. Even the efforts of writing to produce fixity—can these be read as attempts to ensure the survival of the natural, to sustain its cathexis? What the *Riwaj* did was make a primitivist gesture: it designated certain practices as belonging to a time apart from and before colonial rationality, and it tried to bring them into the present constituted by colonial rationality by codifying them, in what it thought was their irreducibility, within colonial rationality.

This is the sense in which colonial writing or the extension work of development regimes are the technologies of primitivism. Contrary to what we usually think, these are not technologies designed to produce transparency and legibility; rather, they should be read as part of the acknowledgment, maybe even production, of opacity, of the failure of sight, of resistance to transcription. Primitivism, with its production of the natural, is in this sense not simply that which it is claimed to be (the affirmation of the superseded past in the modern present). It gestures, rather, to the constitutive outside, both aporetic and enabling, of the practices around the extension of the power of development regimes; it signals constantly to that which development regimes may produce but cannot encompass.

FOUR

Primitivism takes on new forms in postcolonial development regimes; now cathexes of the natural become especially intriguing because of the complex relations between the popular and the nation-state. India's development regime shares that distinctive ventriloquism of modernity where the nation-state has to speak not simply in the name of the popular or for the popular but *as* the popular. That is to say, the nation-state is presented and understood as the embodiment of the people. There are two operations involved in this metonym: an identification of the popular with the national, and an understanding of the nation as realizing itself principally in the form of the nation-state. At meetings that Nehru addressed during the independence struggle—the struggle, in a sense, for the Indian nation to realize itself as a nation-state—he would explain to crowds that India was not so much lands, rivers, and forests as "the people of India." "You are part of this *Bharat Mata,* I told them, you are in a manner yourselves Bharat Mata, and as the idea slowly soaked into their brains, their eyes would light up as though they had made a great discovery" (Nehru [1946] 1981, 60).

And yet this is a peculiar metonym, for it is haunted by the anxiety that the nation-state and the popular are fundamentally different entities. The popular is repeatedly figured in modernist thought as natural. To recall Ranajit Guha's (1983) analysis of the colonial discourse of counterinsurgency, popular rebellion is repeatedly understood through analogies with natural phenomena: rebellions spread, for example, like wildfire. This association of the popular and the natural has persisted as a significant trope in nationalist thought—recall Nehru's lyrical evocations in his writings of the Indian peasant and his deep connections with the land.

In contrast, the Indian nation-state, despite its claims to identity with the popular, was not only thought of as natural; in the claims to history that were made from almost the inception of nationalist thought, the nation-state was to be precisely that which was not natural. To become a nation-state was, for Indian nationalists, to become capable of growth, progress, and modernity. This, then, is the paradox. The Indian nation-state is marked by a yearning for the popular, by the desire to claim a metonymic relationship or even an identity with the national-popular, and simultaneously by an awareness of the impossibility of this yearning, of the difference between it and the popular. Therefore, the form in which the nation-state relates to the popular is through primitivism—that is to say, through a yoking together of two different times, through an always anachronistic staging of the popular in the time of the nation-state. In this new sense, primitivism is constitutive of the Indian nation-state, too.

All of this has implications for the ways in which we think of the Indian development regimes. Development cannot only be thought of in the ways it represents itself: as a force of history, as an attempt to transform the natural-popular into the historical by bringing about modernity and progress. Nor can it be thought of in terms of its instrument effects of extending bureaucratic power, of depoliticizing poverty by claiming that it has technical solutions. Development is all of that, of course. But it is also more. It is a primitivism: it stages the natural-popular within the historical and thus attempts a closure; it is a politics that attempts to sustain the nation-state by suspending its perceived lacks—the lack both of the popular and of progress. That is to say, the popular is not only the object of development; it is also that which development would like to both erase and identify with.

FIVE

In this context, it is a profound irony that so much of the environmentalism that has opposed the nation-state's development has also been primitivist; it has been an affirmation of the natural against the vicissitudes of modernity. It was quite recently, after all, that Max Oelschlaeger (1991), speaking in the name of postmodern sensibilities, no less, called for a "new primitivism," one that would celebrate the natural without being violent toward it. For such environmentalism, opposition to the nation-state's projects springs from the apprehension that it is not primitivist enough; this environmentalism calls on the nation-state to be more committed to producing and affirming the natural.

If we were to understand development regimes as being only about the extension of power (and not sense that they are about the production of resistance and opacity), then environmentalists such as Oelschlaeger would have to be seen as opposed to development regimes. Certainly, such environmentalists are often hostile to the rationality of the state, to most development projects. They trace this rationality often to a modern Western worldview that insists on subordinating nature, or has an instrumental view of nature, a worldview that is seen as having its origins variously in Christianity, the Renaissance, or the Enlightenment. They contrast this, depending on their concerns, with women, tribes, or precolonial, premodern, or traditional peoples, who are ascribed a more sacred, more reverential, noninstrumental view of nature. This view is not entirely incorrect, at least not in the genealogies it provides of instrumental views of nature. Still, there is a certain irony to its logic. After all, the idea of a divine element residing in nature rarely received such a strong affirmation as it did in the post-Renaissance period. Indeed, it is a secularized version of this perspective that informs primitivist environmentalism and its cathexis of the natural.

Given all this, is it possible to see primitivist environmentalism as radical? In a sense, of course, yes. However much we may criticize Gadgil, Guha, and Shiva, we are certainly more likely to find them allied against the Indian state than with it; a liberal postmodernism would be far more collusive with dominant groups than they would be. Still, we must remember that in its opposition to development, primitivist environmentalism shares the same framework as development, even if its emphases are different, even if it celebrates the primitive more emphatically. Primitivism affirms evolutionary and linear time; it simply inverts the valences of that time, placing at the cutting edge of modernity not that which is contemporary but that which is supposed to have been left behind. And such inversion is not enough; what we may need to do is to abandon the framework itself, to move away from the modernist cathexis of nature.

SIX

What could be involved in such a move? A turn to history? That has seemed an increasingly attractive option in the last few decades. It seems the obvious move, the perfect dismantling of the opposition between the natural and the historical. But I hesitate. Rehearsing some background may help situate this hesitancy. In his deeply flawed but brilliant book *The Idea of Nature*, R. G. Collingwood identified three cosmological movements in Western understandings of nature. Greek natural science regarded the presence of mind in nature as the source of regularity and orderliness; for Greek thinkers, the world of nature was not only alive but intelligent, and in this sense it shared a psychic and intellectual kinship with humans. The second cosmological movement was post-Renaissance, associated among others with Galileo, Newton, and Kant. It denied that the world of nature was an organism and insisted that devoid of intelligence and life, it was incapable of ordering its movements in a rational manner. Nature was now viewed as a machine, "an arrangement of bodily parts designed and put together and set going by an intelligent mind outside itself" (Collingwood 1946, 5). As such a machine, it was governed by ascertainable laws. The third cosmological move—the one that Collingwood embraced—identified nature with history. Of course, his understanding of the convergence between nature and history was a somewhat evolutionist and historicist one, beginning with his identification of the shift to a historical idea of nature with Darwin. Still, I think his point survives historicism and evolutionism; there has certainly been a shift to a historical understanding of nature. It is salutary to remind ourselves that the discipline we sometimes claim allegiance to—environmental history—is itself made possible by the seizure of the natural by the historical.

But what troubles me about the historical idea of nature is this. At the head of *The Political Unconscious*, Fredric Jameson (1989) chose to put the motto "Always historicize," apparently quite unaware of the paradoxes of that formulation. It is precisely this "always" that we need to interrogate. Why does claiming agency involve claiming history? Why should identity, personhood, and agency seem unthinkable without history? Where does this refusal of imagination spring from? Why do we refuse to think of other forms of personhood, identity, or agency? Perhaps because history itself is a constitutive myth of modernity (see Skaria 1999). This myth informs the modernist cathexis of the natural; it makes possible the distinction between the historical and the natural. In that sense, when we do environmental history, when we affirm a historical idea of nature but do not question the idea of the historical, we resort to a peculiar modernist cathexis of history, one that has as its inevitable double the modernist cathexis of the natural.

If not the convergence of history and nature, if not a historical idea of nature, then what? Beyond history? This notion of going beyond is informed by some notion of synthesis and transcendence that I find unhelpful. My hope lies elsewhere, in the kinds of themes and issues raised by Cecile Jackson, Molly Chattopadhyah, Shubhra Gururani, and Paul Robbins, among others. What I find most exciting about these chapters is what might be called the supplementarity of the subaltern that characterizes them, and the incoherence of the modernist cathexis of nature in the face of this. I do not mean this simply in the sense that the subaltern groups on whom these papers focus refuse to cathect the natural in the ways characteristic of the hyperreal Europe, that the natural disappears from sight in a welter of practices—that may be the case to some extent, but it scarcely needs to be the case in other situations.

I refer, rather, to the fact that these subaltern practices can be seen as moments of excess in relation to the development regime's cathexis of nature. That is to say, they engage with this cathexis but exceed it; they convert the Indian development regime's history into hybrid histories. These hybrid histories are the supplement that development regimes try to stage as the natural and the primitive, which they try to subordinate to the historical; to foreground instead the supplementarity is to render problematic both the historical and the natural. And yet, not quite—is it not historical to render historical the historical itself?

Bibliography

Abel, Wilhelm. 1974. *Massenarmut und Hungerkrisen im vorindustriellen Europa: Versuch einer Synopsis.* Hamburg: Parey.

Abu-Lughod, J. 1989. *Before European Hegemony: The World System A.D. 1250–1350.* New York: Oxford University Press.

Adas, Michael. 1989. *Machines as the Measure of Man.* Ithaca, N.Y.: Cornell University Press.

Adnan, Shapan. 1998. Fertility Decline under Absolute Poverty: Paradoxical Aspects of Demographic Change in Bangladesh. *Economic and Political Weekly* 33, no. 22 (30 May).

Agarwal, Anil, and Sunita Narain. 1992. *Towards a Green World: Should Global Environmental Management Be Built on Legal Conventions of Human Rights?* New Delhi: Center for Science and Environment.

Agarwal, Bina. 1988. Who Sows? Who Reaps? Women and Land Rights in India. *Journal of Peasant Studies* 15, no. 4: 531–81.

———. 1992. The Gender and Environment Debate: Lessons from India. *Feminist Studies* (spring): 119–58.

———. 1994. *A Field of One's Own: Gender and Land Rights in South Asia.* Cambridge: Cambridge University Press.

Agarwal, V. P. 1988. Wasteland Afforestation in India. In *Wasteland Development for Fuelwood and Fodder Production,* 171–79. Dehra Dun: Forest Research Institute and Colleges, Government of India.

Aggers, Ben. 1992. *The Discourse of Domination: From Frankfurt School to Postmodernism.* Evanston: Northwestern University Press.

Agnihotri, Indu. 1996. Ecology, Land Use, and Colonization: The Canal Colonies of Punjab. *Indian Economic and Social History Review* 33, no. 1: 37–58.

Agrawal, Arun. 1992. The Grass Is Always Greener on the Other Side: A Study of the Raikas, Migrant Pastoralists of Rajasthan. Drylands Networks Programme Issues. Paper no. 36. London: International Institute for Environment and Development.

———. 1993. Mobility and Cooperation among Nomadic Shepherds: The Case of the Raikas. *Human Ecology* 21, no. 3: 261–79.

———. 1994a. I Don't Need It but You Can't Have It: Politics on the Commons. Pastoral Development Network Paper 36a. London: Overseas Development Institute.

———. 1994b. Mobility and Control among Nomadic Shepherds: The Case of the Raikas II. *Human Ecology* 22, no. 2: 131–44.

———. 1994c. Rules, Rule Making, and Rule Breaking: Examining the Fit between Rule Systems and Resource Use. In *Rules, Games, and Common Pool Resources,* ed. Elinor Ostrom, Roy Gardner, and James Walker. Ann Arbor: University of Michigan Press.

———. 1995. Dismantling the Divide between Indigenous and Scientific Knowledge. *Development and Change* 26: 413–39.

———. 1996a. Forest Management under Common Property Regimes in Kumaon Himalaya. Paper prepared for the Conference "Participation, People, and Sustainable Development: Understanding the

Dynamic of Natural Resource Systems."
Tribhuvan University, Nepal.

———. 1996b. Group Size and Collective
Action: A Case Study of Forest Manage-
ment Institutions in the Indian Himalaya.
In Forest, Trees, and People Programme.
Phase 2. Working Paper no. 3. Rome.
FAO.

———. 1997. Community in Conservation:
Beyond Enchantment and Disenchant-
ment. Working Paper. Gainesville, Fla.:
Conservation and Development Forum,
University of Florida.

———. 1999. *Greener Pastures: Markets, Poli-
tics, and Community among a Migrant
Pastoral People.* Durham, N.C.: Duke
University Press.

Agrawal, Arun, and Gautam Yadama. 1997.
How Do Local Institutions Mediate
Market and Population Pressures on Re-
sources: Forest *Panchayats* in Kumaon.
Development and Change 28: 435–65.

Ahuja, Kanta, and M. S. Rathore. 1987.
Goats and Goatkeepers. Jaipur: Institute of
Development Studies.

Alavi, H. 1980. India: Transition from Feu-
dalism to Colonial Capitalism. *Journal of
Contemporary Asia* 10, no. 4: 359–99.

Alderman, Harold, and Christine Paxson.
1992. Do the Poor Insure? A Synthesis of
the Literature on Risk and Consumption
in Developing Countries. *World Bank
Policy Research Working Paper,* WPS 1008.
Washington, D.C.: World Bank.

Anderson, Alexander. 1887. *Report on the Forest Settlement in the Kangra Valley.* Lahore:
Punjab Government Press.

———. 1897. *Final Report of the Revised Settlement of Kangra Proper.* Lahore: Civil and
Military Gazette Press.

Anderson, David. 1984. Depression, Dust
Bowl, Demography, and Drought: The
Colonial State and Soil Conservation

in East Africa during the 1930s. *African
Affairs* 83: 321–43.

Andranovich, Greg, and Nicholas P. Lovrich.
1996. Editor's Introduction: Community-
Oriented Research. (Community-
Oriented Research: Grassroots Issues
versus National Policy Agendas.) *American Behavioral Scientist* 39, no. 5 (March–
April): 525–36.

Annual Progress Reports of the Punjab
Forest Department. 1904–1914.

Appadurai, Arjun. 1996. *Modernity at Large:
Cultural Dimensions of Globalization.*
Minneapolis: University of Minnesota
Press.

Apte, D. V. 1920. *Chandrachud Daftar.* Pune:
Bharata Itihasa Samshodhaka Mandala.

Arizpe, Lourdes, M. Priscilla Stone, and
David C. Major, eds. 1994. *Population
and Environment: Rethinking the Debate.*
Boulder: Westview Press.

Arnold, David. 1993. *Colonizing the Body:
State Medicine and Epidemic Disease in
Nineteenth-Century India.* Berkeley:
University of California Press.

———. 1996. *The Problem of Nature: Environment, Culture, and European Expansion.*
Oxford: Blackwell.

Arnold, David, and Ramachandra Guha.
1995. *Nature, Culture, and Imperialism:
Essays on the Environmental History of South
Asia.* Delhi: Oxford University Press.

Arnold, J. E. M. 1990. Common Property
Management and Sustainable Devel-
opment in India. Working Paper no. 9,
Forestry for Sustainable Development
Program. Minneapolis: University of
Minnesota.

Arnold, J. E. M., and W. C. Stewart. 1991.
Common Property Resource Management in India. Tropical Forestry Papers
no. 24, Oxford Forestry Institute. Oxford:
Oxford University.

Arrighi, G. 1994. *The Long Twentieth Century: Money, Power, and the Origins of Our Times.* London: Verso.

Ashish, Madhav. 1983. Agricultural Economy of Kumaon Hills: Threat of Ecological Disaster. In *The Himalayas: Nature, Man, Culture.* New Delhi: Rajesh Publications.

———. 1993. Legal Restrictions on Village Powers of Land Management as a Major Cause of Land Degradation in the UP Hills. *CHEA Bulletin* 6: 1–11.

Atkinson, Adrian. 1991. *Principles of Ecology.* London: Bellhaven.

Atkinson, E. T. 1882a. *The Himalayan Gazetteer.* Vol. 3, part 1. Reprint, 1989. New Delhi: Cosmo Publications.

———. 1882b. *The Himalayan Gazetteer.* Vol. 1, Part 2. Reprint, 1989. New Delhi: Cosmo Publications.

Atre, Tryambak N. 1915. *Gaon-Gada.* Reprint, 1989. Pune: Varda Books.

Baden-Powell, B. H. 1892. *Land Systems in British India.* Oxford: Clarendon Press.

Bahuguna, S. 1982. Let the Himalayan Forests Live. *Science Today,* March, 41–46.

Ballabh, Vishwa, and Kartar Singh. 1988. Van (Forest) Panchayats in Uttar Pradesh Hills: A Critical Analysis. Research Paper No. 2. Institute of Management. Anand.

Ballhatchet, K. A. 1957. *Social Policy and Social Change in Western India, 1817–1830.* London.

Bandyopadhyay, J. 1983. The Challenges of Social Forestry. In *Towards a New Forest Policy: People's Rights and Environmental Needs.* New Delhi: Indian Social Institute.

Banuri, T., and F. Apffel-Marglin. 1993. A Systems-of-Knowledge Analysis of Deforestation. In *Who Will Save the Forests: Knowledge, Power, and Environmental Destruction,* ed. T. Banuri and F. Apffel-Marglin, 1–23. London: Zed Books.

Barnes, George Carnac. 1855. *Report of the Land Revenue Settlement of the Kangra District, Punjab.* Lahore: Civil and Military Gazette Press.

Barth, Frederick. 1961. *The Nomads of South Persia.* Boston: Little Brown.

———. 1964. Capital, Investment, and the Social Structure of a Pastoral Nomad Group in South Persia. In *Capital, Saving, and Credit in Peasant Societies,* ed. R. Firth and B. S. Yamey, 69–81. Chicago: Aldine Publishing.

Bates, Crispin. 1981. The Nature of Social Change in Rural Gujarat: The Kheda District, 1818–1918. *Modern Asian Studies* 15, no. 4.

Baviskar, Amita. 1995. *In the Belly of the River: Tribal Conflicts over Development in the Narmada Valley.* Delhi: Oxford University Press.

Bayly, Chris. 1983. *Rulers, Townsmen, and Bazaars.* Cambridge: Cambridge University Press.

Behnke, Roy, and Ian Scoones. 1993. *Rethinking Range Ecology: Implications for Rangeland Management in Africa.* London: International Institute for Environment and Development.

Beinart, William. 1984. Soil Erosion, Conservationism, and Ideas about Development: A Southern African Exploration, 1900–1960. *Journal of Southern African Studies* 11, no. 1: 52–83.

———. 1989. The Politics of Colonial Conservation. *Journal of Southern African Studies* 15, no. 2: 143–62.

Benton, Ted, ed. 1996. *The Greening of Marxism.* New York: Guilford Press.

Berkes, Fikret, ed. 1989. *Common Property Resources: Ecology and Community-Based Sustainable Development.* London: Belhaven Press.

Berreman, G. 1989. Chipko: A Movement to Save the Himalayan Environment and

People. In *Contemporary Indian Tradition: Voices on Culture, Nature, and the Challenge of Change,* ed. C. M. Borden, 239–66. Washington, D.C.: Smithsonian.

Berry, Sara. 1988. Concentration without Privatization? Some Consequences of Changing Patterns of Rural Land Control in Africa. In *Land and Society in Contemporary Africa,* ed. R. E. Downs and S. P. Reyna, 53–75. Hanover, N.H.: University Press of New England.

Béteille, Andre. 1965. *Caste, Class, and Power: Changing Patterns of Stratification in a Tanjore Village.* Berkeley: University of California Press.

Bhaskar, Roy. 1989. *Reclaiming Reality.* London: Verso.

———. 1993. *Dialectic: The Pulse of Freedom.* London: Verso.

Bhatia, B. 1997. Reasserting Dominance in the Face of Resistance: Caste Senas and the Naxalite Movement in Central Bihar. Paper given to Workshop on Rural Labour Relations in India Today. London School of Economics, 19–20 June.

Bhatt, C. P. 1987. The Chipko Movement: Strategies, Achievements, and Impacts. In *The Himalayan Heritage,* ed. M. K. Raha, 238–48. New Delhi: Gian Publishing House.

Bhattacharya, Neeladri. 1995. Pastoralists in a Colonial World. In *Nature, Culture, Imperialism: Essays on the Environmental History of South Asia,* ed. D. Arnold and R. Guha. Delhi: Oxford University Press.

———. 1998. Introduction. *Studies in History,* n.s., 14, no. 2: 165–71.

Binns, Tony. 1990. Is Desertification a Myth? *Geography* 75: 106–13.

Binswanger, Hans. 1980. *The Economics of Tractorization.* Delhi: Oxford University Press.

Blaikie, Piers. 1985. *Political Economy of Social Erosion in Developing Countries.* London: Longman.

Blaikie, Piers, and Harold Brookfield. 1987. *Land Degradation and Society.* New York: Methuen.

Blyn, George. 1966. *Agricultural Trends in India, 1891–1947: Output, Availability, and Productivity.* Philadelphia: University of Pennsylvania Press.

Bonner, Arthur. 1990. *Averting the Apocalypse: Social Movements in India Today.* Durham, N.C.: Duke University Press.

Bormann, F. Herbert, and Gene E. Likens. 1979. *Pattern and Process in a Forested Ecosystem.* New York: Springer-Verlag.

Bose, A. B. 1975. Pastoral Nomadism in India: Nature, Problems, and Prospects. In *Pastoralists and Nomads in South Asia,* ed. L. S. Leshnik and G.-D. Sontheimer. Wiesbaden: Otto Harrassowitz.

Bourdieu, Pierre. 1977. *Outline of a Theory of Practice.* Trans. Richard Nice. Cambridge: Cambridge University Press.

———. 1990. *The Logic of Practice.* Trans. Richard Nice. Stanford: Stanford University Press.

Brandis, D. 1897. *Forestry in India: Origins and Early Development.* Reprint, 1994. Dehra Dun: Natraj Publishers.

Brara, Rita. 1987. *Shifting Sands: A Study of Rights in Common Pastures.* Jaipur: Institute for Development Studies.

Braudel, F. 1977. *Afterthoughts on Material Civilization and Capitalism.* Baltimore: Johns Hopkins University Press.

———. 1982. *The Wheels of Commerce.* New York: Harper and Row.

———. 1984. *The Perspective of the World.* New York: Harper and Row.

Braun, Bruce, and Noel Castree, eds. 1998. *Remaking Reality: Nature at the Millennium.* London: Routledge.

Brokensha, David, D. Michael Warren, and

O. Werner. 1980. *Indigenous Knowledge Systems in Development.* Lanham, Md.: University Press of America.

Bromley, David. 1989. Property Relations and Economic Development: The Other Land Reform. *World Development* 17, no. 6: 867–77.

———, ed. 1992. *Making the Commons Work: Theory, Practice, and Policy.* San Francisco: ICS Press.

Brown, Leslie H. 1971. The Biology of Pastoral Man as a Factor in Conservation. *Biological Conservation* 3: 93–100.

Brush, Stephen, and Doreen Stabinksy, eds. 1996. *Valuing Local Knowledge: Indigenous People and Intellectual Property Rights.* Washington, D.C.: Island Press.

Bryant, Raymond. 1992. Political Ecology: An Emerging Research Agenda in Third-World Studies. *Political Geography* 11, no. 1: 12–36.

Busch, J. M., and J. D. Hewlett. 1982. A Review of Catchment Experiments to Determine the Effect of Vegetation Changes on Water Yield and Evapotranspiration. *Journal of Hydrology* 55: 3–23.

Buttel, Frederick. 1992. Environmentalization: Origins, Processes, and Implications for Rural Third-World Social Change. *Rural Sociology* 57, no. 1: 1–27.

———. 1996. Environmental and Resource Sociology: Theoretical Issues and Opportunities for Synthesis. *Rural Sociology* 61: 56–76.

Byres, Terry. 1981. The New Technology, Class Formation, and Class Action in the Indian Countryside. *Journal of Peasant Studies* 8, no. 4.

Cautley, P. T. 1860. *Report on the Ganges Canal Works: From Their Commencement until the Opening of the Canal in 1854.* Vol. 1. London: Smith, Elder.

Cernea, Michael. 1981. Sociological Dimensions of Extension Organization: The Introduction of the T&V System in India. In *Extension Education and Rural Development,* ed. Bruce R. Crouch and Shankariah Chamala. Chichester, U.K.: Wiley.

Chakrabarty, Dipesh. 1992. Postcoloniality and the Artifice of History: Who Speaks for the "Indian" Pasts. *Representations* 37 (winter).

Chakravarty-Kaul, Minoti. 1996. *Common Lands and Customary Law: Institutional Change in North India over the Past Two Centuries.* Delhi: Oxford University Press.

Chambers, Robert, N. C. Saxena, and T. Shah. 1989. *To the Hands of the Poor: Water and Trees.* New Delhi: Oxford and IBH Publishing.

Champion, H. G., and F. C. Osmaston. 1962. *E. P. Stebbing's "The Forests of India," Being the History from 1925 to 1947 of the Forests Now in Burma, India, and Pakistan.* Vol. 4. Reprint, 1983. Delhi: Periodical Expert Book Agency.

Chatterjee, Partha. 1982. Agrarian Relations and Communalism in Bengal, 1926–1935. In *Subaltern Studies: Writings on South Asian History and Society,* vol. 1, ed. Ranajit Guha, 9–38. Delhi: Oxford University Press.

———. 1983. More on Modes of Power and the Peasantry. In *Subaltern Studies II: Writings on South Asian History and Society,* ed. Ranajit Guha, 311–49. Delhi: Oxford University Press.

———. 1984. Gandhi and the Critique of Civil Society. In *Subaltern Studies III: Writings in South Asian History and Society,* ed. Ranajit Guha, 153–95. Delhi: Oxford University Press.

———. 1993. *The Nation and Its Fragments.* Delhi: Oxford University Press.

Chattopadhyaya, Brijadulal. 1994. *The*

Making of Early Medieval India. Delhi: Oxford University Press.

Chayanov, A. V. [1966] 1986. *The Theory of Peasant Economy*. Madison: University of Wisconsin Press.

Chen, Martha A. 1993. Women and Wasteland Development in India: An Issue Paper. In *Women and Wasteland Development in India,* ed. A. M. Singh and N. Burra. New Delhi: Sage Publications.

Chopra, K., G. K. Kadekodi, and M. N. Murty. 1990. *Participatory Development: People and Common Property Resources*. New Delhi: Sage Publications.

Chua, Cathy. 1986. The Development of Capitalism in Indian Agriculture: Gujarat, 1850–1900. *Economic and Political Weekly* 21.

Cincotta, Richard, and G. Pangare. 1994. Population Growth, Agricultural Change, and Natural Resource Transition: Pastoralism amidst the Agricultural Economy of Gujarat. Pastoral Development Network Paper 36a. London: Overseas Development Institute.

Clark, Alice. 1979. Central Gujarat in the Nineteenth Century: The Integration of an Agrarian System. Ph.D. diss., University of Wisconsin, Madison.

———. 1983. Limitations on Female Life Chances in Central Rural Gujarat. *Indian Economic and Social History Review* 20, no. 1.

Cleghorn, H., F. Royle, H. Baird Smith, and R. Strachey. 1851. *To Consider the Probable Effects in an Economical and Physical Point of View of the Destruction of Tropical Forests*. Edinburgh: British Association Report.

Cohn, Bernard S. 1987. The Initial British Impact on India: A Case Study of the Benares Region. In *An Anthropologist among the Historians and Other Essays*. Delhi: Oxford University Press.

Coleman, J. 1988. Social Capital in the Creation of Human Capital. *American Journal of Sociology* 94: S95–S120.

Collectorate Records, Pre-Mutiny. 1816–1857. Papers regarding the Cultivation of Hemp in Garhwal. Serial no. 89. File no. 35. Dehra Dun: U.P. State Regional Archives.

Collingwood, R. G. 1946. *The Idea of Nature*. Oxford: Oxford University Press.

Comaroff, Jean. 1985. *Body of Power, Spirit of Resistance*. Chicago: University of Chicago Press.

Comaroff, Jean, and John Comaroff. 1991. *Of Revelation and Revolution: Christianity, Colonialism, and Consciousness in South Africa*. Vol. 1. Chicago: University of Chicago Press.

Commons, John R. 1990. *Institutional Economics: Its Place in Political Economy*. New Brunswick: Transaction Publishers.

Connolly, V. 1911. *Preliminary Assessment Report of the Dehra and Hamirpur Tahsils of the Kangra District*. Lahore: Punjab Government Press.

Cosgrove, Denis, and Stephen Daniels, eds. 1988. *The Iconography of Landscape*. Cambridge: Cambridge University Press.

Cronon, William. 1991. *Nature's Metropolis: Chicago and the Great West*. New York: W. W. Norton.

———, ed. 1995. *Uncommon Ground: Toward Reinventing Nature*. New York: W. W. Norton.

CSE. 1985. The State of India's Environment: The Second Citizen's Report. Reprint 1986. New Delhi: Centre for Science and Environment.

———. 1991. Floods, Flood Plains, and Environmental Myths. State of India's Environment: Third Citizen's Report. New Delhi: Centre for Science and Environment.

Dahl, G. 1987. Women in Pastoral Production. *Ethnos* 52, nos. 1–2: 246–77.

Damodaran, Vinita. 1995. Famine in a Forest Tract: Ecological Changes and the Causes of the 1897 Famine in Chotanagpur, Northern India. *Environment and History* 1, no. 2: 129–58.

Dangwal, Dhirendra Datt. 1998. Forests, Farms, and Peasants: Agrarian Economy and Ecological Change in the U.P. Hills, 1815–1914. *Studies in History*, n.s., 14, no. 2: 349–71.

Davison, Robin. 1993. Wandering with India's Rabari. *National Geographic* 184, no. 3: 64–93.

Deere, Carmen Diana, and Alain de Janvry. 1981. Demographic and Social Differentiation among Northern Peruvian Peasants. *Journal of Peasant Studies* 8.

de Janvry, Alain. 1981. *The Agrarian Question and Reformism in Latin America*. Baltimore: Johns Hopkins University Press.

Deleuze, Gilles, and Félix Guattari. 1988. *A Thousand Plateaus: Capitalism and Schizophrenia*. Trans. Brian Massumi. London: Athlone Press.

Demeritt, David. 1998. Science, Social Constructivism, and Nature. In *Remaking Reality: Nature at the Millennium,* ed. B. Braun and N. Castree, 173–93. London: Routledge.

den Tuinder, Nico. 1992. Population and Society in Kheda District (India), 1819–1921: A Study of the Economic Context of Demographic Developments. Ph.D. diss., Universiteit van Amsterdam.

Desai, M. B. 1948. *The Rural Economy of Gujarat*. London: Oxford University Press.

Devalle, S. 1992. *Discourses of Ethnicity: Culture and Protest in Jharkhand*. New Delhi: Sage.

Devine, T. M. 1988. *The Great Highland Famine: Hunger, Emigration, and the Scottish Highlands in the Nineteenth Century*. Edinburgh: Donald.

De Waal, Alexander. 1989. *Famine That Kills: Darfur, Sudan, 1984–1985*. Oxford: Oxford University Press.

Dewey, C. 1978. The End of the Imperialism of Free Trade: The Eclipse of the Lancashire Lobby and the Concession of Fiscal Autonomy to India. In *The Imperial Impact,* ed. C. Dewey and A. G. Hopkins, 35–67. University of London, Institute of Commonwealth Studies, Commonwealth Papers 21. London: Athlone Press.

Dhanagare, D. N. 1983. *Peasant Movements in India, 1920–50*. Delhi: Oxford University Press.

Dhiman, D. R. 1988. People's Participation in Wasteland Development. In *Wasteland Development for Fuelwood and Fodder Production,* 131–39. Dehra Dun: Forest Research Institute and Colleges, Government of India.

Dirks, Nicholas B. 1985. Terminology and Taxonomy, Discourse and Domination: From Old Regime to Colonial Regime in South India. In *Studies of South India: An Anthology of Recent Research and Scholarship,* ed. R. Frykenberg and P. Kolenda. Madras: New Era Publications.

———. 1992a. Castes of Mind. *Representations* 37: 56–78.

———. 1992b. From Little King to Landlord: Colonial Discourse and Colonial Rule. In *Colonialism and Culture,* ed. N. B. Dirks. Ann Arbor: University of Michigan Press.

Dogra, B. 1983. *Forests and People: A Report on the Himalayas*. New Delhi: Dogra.

Donham, Donald. 1990. *History, Power, Ideology*. Cambridge: Cambridge University Press.

Douie, J. M. [1899] 1985. *Punjab Settlement Manual*. Delhi: Daya Publishing House.

Dube, S. C. 1969. Social Structure and Change in Indian Peasant Communities. In *Rural Sociology in India*, ed. A. R. Desai, 201–5. Bombay: Popular Prakashan.

Dyson-Hudson, N., and R. Dyson-Hudson. 1982. The Structure of East African Herds and the Future of East African Herders. *Development and Change* 13: 213–38.

Eckersley, Robyn. 1992. *Environmentalism and Political Theory: Towards an Ecocentric Approach*. Albany: State University of New York Press.

Eckholm, Erik P. 1975. The Deterioration of Mountain Environments. *Science* 189: 764–70.

———. 1977. Spreading Deserts—the Hand of Man. Washington, D.C.: Worldwatch Paper no. 13.

Economic and Political Weekly. 1992. Development for Whom? Critique of Rajasthan Programme. *Economic and Political Weekly*, 1 February, 193–98.

Elliott, H. M. 1966. *The History of India, as Told by Its Own Historians*. The Muhammadan Period. New York: AMS Press.

Ellis, Frank. 1988. *Peasant Economics*. Cambridge: Cambridge University Press.

Ellis, Jim E., and David M. Swift. 1988. Stability of African Pastoral Ecosystems: Alternate Paradigms and Implications for Development. *Journal of Range Management* 41: 450–59.

Embree, Ainslie T. 1969. Landholding in India and British Institutions. In *Land Control and Social Structure in Indian History*, ed. R. E. Frykenberg. Madison: University of Wisconsin Press.

Engel, J., and J. Engel, eds. 1994. *Ethics of Environment and Development: Global Challenges and International Response*. London: Belhaven Press.

Enthoven, R. E. 1920. *Tribes and Castes of Bombay*. 3 vols. Bombay.

Escobar, Arturo. 1992. Imaging a Post-development Era? Critical Thought, Development, and Social Movements. *Social Text* 23–31: 20–56.

———. 1995. *Encountering Development: The Making and Unmaking of the Third World*. Princeton: Princeton University Press.

———. 1996. Constructing Nature: Elements for a Poststructuralist Political Ecology. In *Liberation Ecologies: Environment, Development, Social Movements*, ed. R. Peet and M. Watts, 46–68. London: Routledge.

Esteva, Gustavo. 1993. Development. In *The Development Dictionary: A Guide to Knowledge as Power*, ed. W. Sachs, 6–25. London: Zed Books.

Esty, Daniel, and Marian Chertow, eds. 1997. *Thinking Ecologically: The Next Generation of Environmental Policy*. New Haven: Yale University Press.

Eswaran, Mukesh, and Ashok Kotwal. 1990. Implications of Credit Constraints for Risk Behaviour in Less-Developed Economies. *Economic Papers* 42.

Evans, P. 1995. *Embedded Autonomy: States and Industrial Transformation*. Princeton: Princeton University Press.

Fairhead, James, and Melissa Leach. 1996. *Misreading the African Landscape: Society and Ecology in a Forest Savannah Mosaic*. Cambridge: Cambridge University Press.

Feder, Gershon, and David Feeny. 1991. Land Tenure and Property Rights: Theory and Implications for Development Policy. *The World Bank Economic Review* 5, no. 1: 135–53.

Felski, Rita. 1995. *The Gender of Modernity*. Cambridge: Harvard University Press.

Ferguson, James. 1990. *The Anti-politics Machine: "Development," Depoliticization, and Bureaucratic Power in Lesotho.* Cambridge: Cambridge University Press.

Ferishtah, Muhammad Qasim Hindu Shah Astrabadi. [1792] 1963. *The History of Hindostan.* Translated from the Persian by Alexander Dow. London: John Murray. Reprint, New Delhi.

Fernandes, Leela. 1997. *Producing Workers: The Politics of Gender, Class, and Culture in the Calcutta Jute Mills.* Philadelphia: University of Pennsylvania Press.

Fernandes, Walter, and Geeta Menon. 1987. *Tribal Women and Forest Economy: Deforestation, Exploitation, and Status Change.* Delhi: Indian Social Institute.

Forest Act. *The Indian Economic and Social History Review* 27, no. 1: 65–84.

Forest Manual 1936. United Province of Agra and Oudh. 6th ed. Allahabad, India: United Provinces.

Forse, B. 1989. The Myth of the Marching Desert. *New Scientist* 4: 31–32.

Foucault, Michel. 1979. *Discipline and Punish: The Birth of the Prison.* New York: Vintage Books.

———. 1980. *Power/Knowledge: Selected Interviews and Other Writings, 1972–1977.* Ed. C. Gordon. New York: Pantheon Books.

———. 1991. Governmentality. In *The Foucault Effect: Studies in Governmentality, with Two Lectures by and an Interview with Michel Foucault,* ed. G. Burchell, C. Gordon, and P. Miller, 87–104. London: Harvester Wheatsheaf.

———. [1994] 1997. The Birth of Biopolitics. In *Michel Foucault: Ethics, Subjectivity, and Truth,* ed. Paul Rabinow, 73–80. New York: New Press.

Fukazawa, H. 1982. Standard of Living: Maharashtra and the Deccan. *The Cambridge Economic History of India.* Vol. 1. Cambridge: Cambridge University Press.

———. 1991. *The Medieval Deccan: Peasants, Social Systems and States.* Delhi: Oxford University Press.

Gaard, G. 1997. Ecofeminism and Wilderness. *Environmental Ethics* 19: 5–24.

Gadgil, Madhav, and K. C. Malhotra. 1982. Ecology of a Pastoral Caste: Gavli Dhangars of Peninsular India. *Human Ecology* 10, no. 1: 107–43.

———. 1983. Adaptive Significance of the Indian Caste System. *Annals of Human Biology* 10, no. 5: 465–78.

Gadgil, Madhav, and Ramachandra Guha. 1992. *This Fissured Land: An Ecological History of India.* Delhi: Oxford University Press.

———. 1993. *This Fissured Land: An Ecological History of India.* Berkeley: University of California Press.

———. 1995. *Ecology and Equity: The Use and Abuse of Nature in Contemporary India.* London: Routledge.

Gadgil, Madhav, and V. D. Vartak. 1975. Sacred Groves of India: A Plea for Continued Conservation. *Journal of the Bombay Natural History Society* 72, no. 2: 312–20.

———. 1981. The Sacred Groves of Maharashtra: An Inventory. In *Glimpses of Indian Ethnobotany,* ed. S. K. Jain. New Delhi: Oxford and IBH Publishing.

Gaikwad, Khanderav, ed. 1971. *Karvir Sardaranchya Kaifiyati.* Kolhapur: Gaikwad.

Galaty, John G. 1984. Cultural Perspectives on Nomadic Pastoral Societies. *Nomadic Peoples* 16: 15–29.

Galaty, John G., and D. L. Johnson. 1990. Introduction: Pastoral Systems in a Global Perspective. In *The World of Pastoralism,* ed. J. G. Galaty and D. L. Johnson, 1–31. New York: Guilford Press.

Ghosh, A. 1995. *In an Antique Land*. New Delhi: Ravi Dayal.

Ghurye, G. S. 1932. *Caste and Race in India*. London: Kegan Paul.

Gibson, Alexander. 1861. *Forest Reports of the Bombay Presidency for the Years 1856-7—1859-60*. Bombay: Government of Bombay.

Gibson-Graham, J. K. 1996. *The End of Capitalism (As We Knew It): A Feminist Critique of Political Economy*. Oxford: Blackwell.

Gidwani, Vinay K. 1996. Fluid Dynamics: An Essay on Canal Irrigation and the Process of Agrarian Change in Matar Taluka, Kheda District (Gujarat), India. Ph.D. diss., University of California, Berkeley.

———. 2000. The Quest for Distinction: A Reappraisal of the Rural Labor Process in Western India. *Economic Geography*, 76 (2): 165–68.

Gilles, Jere L., and J. Gefu. 1990. Nomads, Ranchers, and the State: The Sociocultural Aspects of Pastoralism. In *The World of Pastoralism*, ed. J. G. Galaty and D. L. Johnson, 99–118. New York: Guilford Press.

GOB (Government of Bombay). 1887. Report of the Bombay Forest Commission. 4 vols. Bombay: Government Central Press.

GOB (Government of Bombay). 1921–1992. *Report on the Forest Administration in the Bombay Presidency for 1921-22*. Bombay: Government Central Press.

GOI (Government of India). 1997. *Approach Paper to the Ninth Five-Year Plan*. New Delhi: Planning Commission.

Gold, Ann. 1998. Authority, Responsibility, and Protection: A Fieldwork Report from Rural North India. Working Paper presented at Forum 1997, Istanbul, Turkey: New Linkages in Conservation and Development Forum.

Gold, Ann G., and B. R. Gujar. 1989. Of Gods, Trees, and Boundaries: Divine Conservation in Rajasthan. *Asian Folklore Studies* 48: 211–29.

Goldman, Michael. 1991. Cultivating Hot Peppers and Water Crisis in India's Desert: Toward a Theory of Understanding Ecological Crisis. *Bulletin of Concerned Asian Scholars* 23, no. 4: 19–29.

Goodman, David, Bernardo Sorj, and John Wilkinson. 1987. *From Farming to Biotechnology: A Theory of Agro-industrial Development*. Oxford: Basil Blackwell.

Goodman, David, and Michael Redclift. 1991. *Refashioning Nature: Food, Ecology, and Culture*. London: Routledge.

Gordon, Colin. 1980. Afterword. In *Power/Knowledge: Selected Interviews and Other Writings 1972-1977,* by Michel Foucault, ed. C. Gordon. New York: Pantheon Books.

———. 1991. Government Rationality: An Introduction. In *The Foucault Effect*, ed. G. Burchell, C. Gordon, and P. Miller, 1–52. London: Harvester Wheatsheaf.

Gordon, Stewart. 1996. Robes of Honour: A "Transactional" Kingly Ceremony. *Indian Economic and Social History Review* 33, no. 3: 225–42.

Greaves, Tom, ed. 1994. *Intellectual Property Rights for Indigenous Peoples: A Sourcebook*. Oklahoma City: Society for Applied Anthropology.

Greenough, Paul. 1982. *Prosperity and Misery in Modern Bengal: The Famine of 1943-1944*. New York: Oxford University Press.

Grieve, J. W. A. 1920. Note on the Economics of Nomadic Grazing as Practiced in Kangra District. *Indian Forester* 28: 333.

Grove, Richard. 1995. *Green Imperialism:*

Colonial Expansion, Tropical Island Edens, and the Origins of Environmentalism, 1600–1860. Cambridge: Cambridge University Press.

Grove, Richard, Vinita Damodaran, and Satpal Sangwan. 1996. *Nature and the Orient: Essays on the Environmental History of South and Southeast Asia.* Delhi: Oxford University Press.

Guha, Ramachandra. 1983. Forestry in British and Post-British India: A Historical Analysis. *Economic and Political Weekly* 18, nos. 45–46.

———. 1986. *Commercial Forestry and Social Conflict in the Indian Himalaya.* Forestry for Development Lecture Series. Berkeley: University of California, Dept. of Forestry.

Guha, Ramachandra. 1989a. Radical American Environmentalism and Wilderness Preservation: A Third World Critique. *Environmental Ethics* 11: 71–83.

———. 1989b. *The Unquiet Woods: Ecological Change and Peasant Resistance in the Indian Himalaya.* New Delhi: Oxford University Press.

———. 1990a. An Early Environmental Debate: The Making of the 1878 Forest Act. *Indian Social and Economic History Review* 27: 65–84.

———. 1990b. Towards a Cross-Cultural Environmental Ethic. *Alternatives* 15: 431–47.

———. 1993. Writing Environmental History in India. *Studies in History* 9, no. 1: 119–29.

———. 1997. The Authoritarian Biologist and the Arrogance of Anti-humanism. *Ecologist* 271: 14–20.

Guha, Ramachandra, and Juan Martinez-Alier. 1997. *Varieties of Environmentalism: Essays North and South.* London: Earthscan.

Guha, Ramachandra, and Madhav Gadgil. 1989. State Forestry and Social Conflict in British India. *Past and Present,* no. 123 (May): 141–77.

Guha, Ranajit. 1962. *A Rule of Property for Bengal: An Essay on the Idea of Permanent Settlement.* Paris: Mouton.

———. 1983. The Prose of Counter-insurgency. In *Subaltern Studies,* ed. Ranajit Guha. Delhi: Oxford University Press.

Guha, Sumit. 1992. Introduction. In *Growth Stagnation or Decline? Agricultural Productivity in British India.* Delhi: Oxford University Press.

———. 1995. An Indian Penal Regime: Maharashtra in the Eighteenth Century. *Past and Present,* no. 147.

———. 1997. Rules, Laws, and Powers: A Perspective from the Past. In *Rules, Laws, and Constitutions: A Perspective for Our Times,* ed. Satish Saberwal. New Delhi: Sage.

Gupta, Akhil. 1995. Blurred Boundaries: The Discourse of Corruption, the Culture of Politics, and the Imagined State. *American Ethnologist* 22, no. 2: 375–402.

Gupta, Akhil, and James Ferguson. 1992. Beyond "Culture": Space, Identity, and the Politics of Difference. *Cultural Anthropology* 7, no. 1: 6–23.

Gururani, Shubhra. 1996. Fuel, Fodder, and Forests: Politics of Forest Use and Abuse in Uttarakhand Himalaya, India. Ph.D. diss., Syracuse University.

Habermas, Jürgen. 1973. *Theory and Practice.* Trans. John Viertel. Boston: Beacon Press.

Habib, Irfan. 1963. *The Agrarian System of Mughal India.* New York: Asia Publishing House.

———. 1982. *An Atlas of Mughal Empire:*

Political and Economic Maps with Notes, Bibliography, and Index. Delhi.

———. 1995. *Essays in Indian History: Towards a Marxist Perception*. Delhi.

Hagen, James R. 1988. Gangetic Fields: An Approach to Agrarian History through Agriculture and the Natural Environment, 1600–1970. Paper delivered at the annual meeting of the Association for Asian Studies.

Hallowell, A. Irving. 1943. The Nature and Function of Property as a Social Institution. *Journal of Legal and Political Sociology* 1:115–38.

Halperin, Rhoda H. 1994. *Cultural Economies: Past and Present*. Austin: University of Texas Press.

Hamilton, Larry S. 1987. What Are the Impacts of Himalayan Deforestation on the Ganges-Brahmaputra Lowlands and Delta? Assumptions and Facts. *Mountain Research and Development* 7: 256–63.

Haraway, Donna. 1991. *Simians, Cyborgs, and Women: The Reinvention of Nature*. London: Routledge.

Hardiman, David. 1981. *Peasant Nationalists of Gujarat: Kheda District, 1917–1934*. Delhi: Oxford University Press.

———, ed. 1993. *Peasant Resistance in India, 1858–1914*. Delhi: Oxford University Press.

Hardin, Garrett, and John Baden, eds. 1977. *Managing the Commons*. San Francisco: W. H. Freeman.

Harrison, Robert Pogue. 1992. *Forests in the Shadow of Civilization*. Chicago: University of Chicago Press.

Harriss-White, Barbara. 1996. *A Political Economy of Agricultural Markets in South India: Masters of the Countryside*. New Delhi.

Harvey, David. 1985. The Geopolitics of Capitalism. In *Social Relations and Spatial Structures*, ed. Derek Gregory and John Urry. London: Macmillan.

———. 1989. *The Condition of Postmodernity*. Oxford: Basil Blackwell.

Hayami, Y., and M. Kikuchi. 1981. *Asian Village Economy at the Crossroads: An Economic Approach to Institutional Change*. Tokyo: University of Tokyo Press.

Haynes, Douglas, and Gyan Prakash, eds. 1992. *Contesting Power: Resistance and Everyday Social Relations in South Asia*. Delhi.

Heber, Reginald. 1828. *Narrative of a Journey through the Upper Provinces of India*. 2 vols. London.

Heckscher, E. F. 1935. *Mercantilism*. Vol. 2. London: G. Allen and Unwin.

Henderson, Carol. 1993. State Administration and the Concepts of Peasants and Sedentary Agricultural Production in the Thar Desert. Paper presented at the South Asian Studies Conference, Madison, Wisconsin.

Herring, Ronald. 1998. Celebrating the Local: Scale and Orthodoxy in Political Ecology. Working Paper for the Workshop on Knowledge and Authority in Nature. Cornell University. April.

Hilborn, Ray, and Donald Ludwig. 1993. The Limits of Applied Ecological Research. *Ecological Applications* 3: 550–52.

Hills: A Critical Analysis. Research Paper no. 2. Anand: Institute of Rural Management.

Hirsch, Eric, and Michael O'Hanlon, eds. 1995. *The Anthropology of Landscape: Perspectives on Space and Place*. Oxford: Clarendon.

Hobley, Mary. 1991. Gender, Class, and Use of Forest Resources: The Case of Nepal. In *Women and the Environment*. Prepared by Annabel Rodda. London: Zed Books.

HPFD. 1993. Himachal Pradesh Forest Statistics. Solan, Himachal Pradesh: H.P. Forest Printing Press.

Hyde, Lewis. 1983. *The Gift: Imagination and the Erotic Life of Property.* New York: Vintage Books.

IFRI (International Forestry Resources and Institutions). 1993. IFRI Data Collection Instruction Manual. Workshop in Political Theory and Policy Analysis. Bloomington, Indiana.

Ilahiane, Hsain. 1993. Common Property, Ethnicity, and Social Exploitation in the Ziz Valley, Southeast Morocco. Department of Anthropology, University of Arizona. Mimeo.

Inden, Robert. 1990. *Imagining India.* Cambridge: Blackwell Publishers.

Islam, Mufakarul M. 1996. *Irrigation, Agriculture, and the Raj: Punjab, 1887–1947.* Delhi.

Ives, Jack D., and Bruno Messerli. 1989. *The Himalayan Dilemma: Reconciling Development and Conservation.* New York: Routledge.

Jackson, Cecile. 1993. Women/Nature or Gender/History: A Critique of Ecofeminist "Development". *Journal of Peasant Studies* 20, no. 3: 389–418.

Jacobs, A. H. 1965. African Pastoralism: Some General Remarks. *Anthropological Quarterly* 38: 144–54.

Jain, L. C. 1994. Panchayats: Arrangements for Scientific, Technical, Management, and Other Expertise Support Systems. In *Decentralization: Panchayats in the Nineties,* ed. A. Mukherjee. New Delhi: Vikas Publishing House.

Jain, M. S. 1994. *Surplus to Subsistence.* Delhi: Agam Kala Prakashan.

Jameson, Dr. 1844. Annual Report to the Government. *Journal of the Agri-Horticultural Society of India,* 28 February.

Jameson, Fredric. 1989. *The Political Uncon-scious: Narrative as a Socially Symbolic Act.* London: Routledge.

Jodha, Narpat S. 1985. Population Growth and Decline of Common Property Resources in Rajasthan, India. *Population and Development Review* 11: 247–64.

———. 1986. Common Property Resources and Rural Poor in Dry Regions of India. *Economic and Political Weekly* 21, no. 27: 1169–81.

Johnson, Douglas L. 1969. The Nature of Nomadism: A Comparative Study of Pastoral Migrations in Southwestern Asia and Northern Africa. University of Chicago, Department of Geography, Research Paper no. 118.

Joshi, A. 1987. *Sheep Wool and Woolen Industry in India.* Bikaner: Agro-Botanical Publishers.

Joshi, Gopa. 1983. Forests and Forest Policy in India. *Social Scientist* 11, no. 1: 43–52.

Kalla, S. D. 1993. Livestock Resources of Rajasthan. In *Natural and Human Resources of Rajasthan,* ed. T. S. Chouhan. Jodhpur: Scientific Publishers.

Kautsky, Karl. [1899] 1990. *The Agrarian Question.* 2 vols. New York: Zvan Publications.

Kavoori, P. S. 1990. Pastoral Transhumance in Western Rajasthan: A Report on the Migratory System of Sheep. Jaipur: Institute of Development Studies.

Keay, J. 1983. *When Men and Mountains Meet: The Explorers of the Western Himalayas, 1820–1875.* London: Century.

Kelkar, G., and D. Nathan. 1991. *Gender and Tribe: Women, Land, and Forests.* London: Zed.

Kessinger, Thomas G. 1974. *Vilyatpur, 1848–1968: Social and Economic Change in a North Indian Village.* Berkeley: University of California Press.

Khanka, S. S. 1988. *Labour Force, Employment,*

and *Unemployment in a Backward Economy: A Study of Kumaon Region in Uttar Pradesh*. Bombay: Himalayan Publishers.

Khazanov, Anatoly M. 1994. *Nomads and the Outside World*. Madison: University of Wisconsin Press.

Kheda Jilla Panchayat. N.d. District Agricultural Profile of Kheda District. Nadiad: Statistical Division.

———. N.d. Jilla ni Ankdakiya Rooprekha, 1989–1990. Nadiad: Statistical Division.

Köhler-Rollefson, Ilse. 1992a. The Raika Dromedary Breeders of Rajasthan: A Pastoral System in Crisis. *Nomadic Peoples* 30: 74–83.

———. 1992b. The Raikas of Western Rajasthan, India. *Pastoral Development Network Newsletter*. London: Overseas Development Institute.

———. 1993. Rejoinder to O. P. Kavoori's Comments on "The Raikas of Western Rajasthan, India." *Pastoral Development Network Newsletter*, no. 35. London: Overseas Development Institute.

———. 1994. Pastoralism in Western India from a Comparative Perspective: Some Comments. Pastoral Development Network Paper 36a. London: Overseas Development Institute.

Köhler-Rollefson, Ilse, and H. S. Rathore. 1998. Camel Milk Marketing in Rajasthan: A Report on the Poverty Alleviation Program for Raike Pastoralists. League for Pastoral Peoples: Sadri (India).

Kohli, Atul. 1990. *Democracy and Its Discontent: India's Growing Crisis of Governability*. Cambridge: Cambridge University Press.

Korten, David, ed. 1986. *Community Management: Asian Experience and Perspectives*. Hartford: West Kumarian Press.

Koster, Harold A., and C. Chang. 1994.

Introduction. In *Pastoralists at the Periphery: Herders in a Capitalist World*, ed. C. Chang and H. A. Koster. Tucson: University of Arizona Press.

Krishnamurthy, J. 1983. The Occupational Structure. In *The Cambridge Economic History of India, II, c.1757–c.1970*, ed. Dharma Kumar. Cambridge.

Kumarappa, J. C. 1931. *Survey of Matar Taluka*. Ahmedabad: Gujarat Vidyapith.

Lamprey, Hugh F. 1975. Report on the Desert Encroachment Reconnaissance in Northern Sudan. Nairobi, United Nations Environment Programme.

———. 1983. Pastoralism Yesterday and Today: The Overgrazing Problem. In *Ecosystems of the World 13: Tropical Savannas*, ed. F. Bouliere, 643–66. Amsterdam: Elsevier Scientific Publishing.

Latour, Bruno. 1987. *Science in Action*. Cambridge: Harvard University Press.

Lawson, P. 1993. *The East India Company: A History*. London: Longman.

Leach, Melissa. 1991. Engendered Environments: Understanding Natural Resource Management in the West African Forest Zone. *Institute of Development Studies Bulletin* 22, no. 4: 17–24.

Leaf, M. J. 1992. Irrigation and Authority in Rajasthan. *Ethnology* 31: 115–32.

Lenin, V. I. 1974. *The Development of Capitalism in Russia*. Moscow: Progress Publishers.

Li, Tania. 1996. Images of Community: Discourse and Strategy in Property Relations. *Development and Change* 27, no. 3: 501–28.

Linkenbach, Antje. 1994. Ecological Movements and the Critique of Development: Agents and Interpreters. In *Thesis Eleven, India and Modernity: Decentering Western Perspectives* 39: 63–85.

Ludden, David. 1985. *Peasant History in South*

India. Princeton: Princeton University Press.

———. 1992. India's Development Regime. In *Colonialism and Culture,* ed. Nicholas Dirks. Ann Arbor: University of Michigan Press.

———. 1994. Agricultural Production and Indian History. In *Agricultural Production and Indian History,* ed. David Ludden. Delhi: Oxford University Press.

———. 1996a. Archaic Formations of Agricultural Knowledge in South India. In *Meanings of Agriculture in South Asia,* ed. Peter Robb. Delhi: Oxford University Press.

———. 1996b. Caste and Political Economy in Early-Modern South India: The Case of Tinnevelly District. In *Institutions and Economic Change in South Asia: Historical and Contemporary Perspectives,* ed. Burton Stein and Sanjay Subrahmanyam. New Delhi: Oxford University Press.

———, ed. 1996c. *Contesting the Nation: Religion, Conflict, and the Politics of Democracy in India.* Philadelphia: University of Pennsylvania Press.

———. 1999. *An Agrarian History of South Asia.* Cambridge: Cambridge University Press.

Lukács, Georg. [1968] 1986. *History and Class Consciousness.* Trans. Rodney Livingstone. Cambridge: MIT Press.

Lyall, James B. 1874. *Report of the Land Revenue Settlement of the Kangra District, Punjab.* Lahore: Central Jail Press.

Mace, R. 1991. Conservation Biology: Overgrazing Overstated. *Nature* 349: 280–81.

Mackenzie, John. 1995. *Orientalism: History, Theory, and the Arts.* Manchester: Manchester University Press.

MacLeod, Roy, and Deepak Kumar. 1995. *Technology and the Raj: Western Technology and Technical Transfers in India, 1700–1947.* New Delhi: Sage.

Macpherson, C. B., ed. 1978. *Property: Mainstream and Critical Positions.* Toronto: Toronto University Press.

Mahapatra, L. K. 1975. Pastoralists and the Modern Indian State. In *Pastoralists and Nomads in South Asia,* ed. L. S. Leshnik and G.-D. Sontheimer, 209–19. Wiesbaden: Otto Harrassowitz.

Major Causes of Land Degradation in the UP Hills. *CHEA Bulletin* 6: 1–11.

Mamdani, M. 1996. *Citizen and Subject: Contemporary Africa and the Legacy of Late Colonialism.* Princeton: Princeton University Press.

Mann, Michael. 1995. Ecological Change in North India: Deforestation and Agrarian Distress in the Ganga Jamna Doab, 1800–1850. *Environment and History,* no. 1: 201–20.

Mann, Susan. 1990. *Agrarian Capitalism in Theory and Practice.* Chapel Hill: University of North Carolina Press.

Mann, Susan, and James Dickinson. 1978. Obstacles to the Development of a Capitalist Agriculture. *Journal of Peasant Studies* 5, no. 4.

Manwaring, A. [1898] 1991. *Marathi Proverbs.* Reprint, New Delhi: Asian Educational Services.

Marglin, S. 1990. Losing Touch: The Cultural Conditions of Worker Accommodation and Resistance. In *Dominating Knowledge: Development, Culture, and Resistance,* ed. F. Apffel-Marglin and S. Apffel-Marglin. Oxford: Clarendon Press.

Marx, Karl. [1844] 1988. *Economic and Philosophic Manuscripts of 1844.* Trans. Martin Milligan. Buffalo, N.Y.: Prometheus Books.

Massey, Doreen. 1994. *Space, Place, and*

Gender. Minneapolis: University of Minnesota Press.

McCann, James. 1995. The Plough and the Forest: Narratives of Deforestation in Ethiopia, 1840–1992. Paper presented at the Program in Agrarian Studies. New Haven, Yale University.

McClelland, J. 1835. *Some Inquiries in the Province of Kemaon Relative to Geology, and Other Branches of Natural Science.* Calcutta: Baptist Mission Press.

McCay, Bonnie, and J. M. Acheson, eds. 1987. *The Question of the Commons: The Culture and Ecology of Communal Resources.* Tucson: University of Arizona Press.

McLane, John R. 1993. *Land and Local Kingship in Eighteenth-Century Bengal.* Cambridge: Cambridge University Press.

Mellor, Mary. 1996. Ecofeminism and Eco-socialism: Dilemmas of Essentialism and Materialism. In *The Greening of Marxism,* ed. Ted Benton, 251–67. New York: Guilford Press.

Merchant, Carolyn. 1980. *The Death of Nature: Women, Ecology, and the Scientific Revolution.* Harper and Row: San Francisco.

Meszáros, István. 1970. *Marx's Theory of Alienation.* London: Merlin Press.

Middleton, L. 1919. *Final Report of the Third Revised Land Revenue Settlement of the Palampur, Kangra, and Nurpur Tahsils of the Kangra District.* Lahore: Government Printing.

Migdal, Joel S., Atul Kohli, and Vivienne Shue, eds. 1994. *State Power and Social Forces: Domination and Transformation in the Third World.* New York: Cambridge University Press.

Mitchell, Timothy. 1991. The Limits of the State: Beyond Statist Approaches and Their Critics. *American Political Science Review* 85, no. 1: 77–96.

Mitchell, W. J. T., ed. 1994. *Landscape and Power.* Chicago: University of Chicago Press.

Moench, Marcus, and Jayant Bandhyopadhyay. 1986. People-Forest Interaction: A Neglected Parameter in Himalayan Forest Management. *Mountain Research and Development* 6, no. 1: 3–16.

Mookerji, R. K. 1919. *Local Government in Ancient India.* Oxford: Clarendon Press.

Moorcroft, W., and G. Trebeck. 1841. *Travels in the Himalayan Provinces of Hindustan and the Panjab from 1819 to 1825.* Vols. 1–2. London: John Murray.

Moore, Donald. 1998. Clear Waters and Muddied Histories: Environmental History and the Politics of Community in Zimbabwe's Eastern Highlands. *Journal of Southern African Studies,* 24 (2): 377–403.

Moosvi, Shireen. 1987. *The Economy of the Mughal Empire, c. 1595: A Statistical Study.* Delhi.

Morash, Christopher. 1995. *Writing the Irish Famine.* Oxford: Oxford University Press.

Moreland, William. 1929. *The Agrarian System of Moslem India: A Historical Essay with Appendices.* London. Reprint, New Delhi.

Morrison, E., and S. M. J. Bass. 1992. What About the People? In *Plantation Politics: Forest Plantations in Development,* ed. C. Sargent and S. Bass, 92–130. London: Earthscan Publications.

MSS Eur D.148. Manuscript held in the Oriental and India Office Collection of the British Library, London.

Mukhia, Harbans. 1993. *Perspectives on Medieval History.* New Delhi.

Myers, Norman. 1986. Environmental Repercussions of Deforestation in the Himalaya. *Journal of World Forest Resource Management* 2: 63–72.

Nagarajan, V. 1996. Towards a Theory of

"Embedded Ecologies" in Hinduism. Paper given to the workshop on Women as "Sacred Custodians" of the Earth, CCCRW, Oxford, 14–15 June.

Naik, T. B. 1974. Social Status in Gujerat. In *Tribe, Caste, and Peasantry*, ed. K. S. Mathur and B. C. Agarwal. Lucknow: Ethnographic and Folk Culture Society.

Nandy, Ashis. 1987. Science, Authoritarianism, and Culture. *Traditions, Tyranny, and Utopias: Essays in the Politics of Awareness*. Delhi: Oxford University Press.

———. 1989. The Political Culture of the Indian State. *Daedalus: Another India* 118, no. 4: 1–26.

Nath, Viswa. 1973. Female Infanticide and the Lewa Kanbis of Gujarat in the Nineteenth Century. *Indian Economic and Social History Review* 10, no. 4.

Neale, Walter C. 1969. Land Is to Rule. In *Land Control and Social Structure in Indian History*, ed. R. E. Frykenberg. Madison: University of Wisconsin Press.

Nehru, Jawaharlal. [1946] 1981. *The Discovery of India*. New Delhi: Oxford University Press.

Netting, Robert. 1993. *Smallholders, Householders: Farm Families and the Ecology of Intensive, Sustainable Agriculture*. Stanford: Stanford University Press.

Neumann, Roderick. 1992. Political Ecology of Wildlife Conservation in the Mt. Meru Area of Northeast Tanzania. *Land Degradation and Rehabilitation* 3: 85–98.

Nichols, Robert. 1997. Settling the Frontier: Land, Law, and Society in the Peshawar Valley, 1500–1900. Dissertation in History, University of Pennsylvania, 1997. Chaps. 1–3.

North, Douglass. 1990. *Institutions, Institutional Change, and Economic Performance*. New York: Cambridge University Press.

O'Brien, E. 1889. *Assessment Report of the Palam Ilaqa, Kangra District*. Lahore: Civil and Military Gazette Press.

———. 1890. *Assessment Report of the Palam Taluqa, Palampur Tahsil, Kangra District, 1890*. Lahore: Caxton Printing Works.

———. 1891a. *Assessment Report of the Taluka Rajgiri in the Palampur Tahsil, of the Kangra District*. Lahore: Civil and Military Gazette Press.

———. 1891b. *Assessment Report of the Taluka Banghal, Palampur Tahsil, in the Kangra District*. Lahore: Civil and Military Gazette Press.

Oelschlaeger, Max. 1991. *The Idea of Wilderness: From Prehistory to the Age of Ecology*. New Haven: Yale University Press.

Ó Gráda, Cormac. 1992. For Irishmen to Forget? Recent Research on the Great Irish Famine. In *Just a Sack of Potatoes? Crisis Experience in European Societies, Past and Present*, ed. Antti Häkkinen. Helsinki: SHS.

Omvedt, Gail. 1978. Towards a Marxist Analysis of Caste. *Social Scientist* 6, no. 11: 70–76.

———. 1993. *Reinventing Revolution: New Social Movements and the Socialist Tradition in India*. Armonk: New York: M. E. Sharpe.

Ortner, Sherry. 1995. Resistance and the Problem of Ethnographic Refusal. *Comparative Studies in Society and History* 37: 173–93.

Ostrom, Elinor. 1990. *Governing the Commons: The Evolution of Institutions for Collective Action*. Cambridge: Cambridge University Press.

———. 1992. *Crafting Institutions for Self-Governing Irrigation Systems*. San Francisco: Institute for Contemporary Studies.

Oturkar, Rajaram V., ed. 1950. *Peshvekalina*

Samajik va Arthik Patravyvahara. Mandala Pune: Bharata Itihasa Samshodhaka.

Pagden, Anthony. 1986. *The Fall of Natural Man: The American Indian and the Origins of Comparative Ethnology.* Cambridge: Cambridge University Press.

Pant, Govind Ballabh. 1922. *The Forest Problem in Kumaon.* Nainital: Gyanodaya Prakashan.

Parmar, B. S. 1959. Report on the Grazing Problems and Policy of Himachal Pradesh. Simla, Himachal Pradesh Forest Department.

Parry, Jonathon P. 1979. *Caste and Kinship in Kangra.* New Delhi: Vikas Publishing House.

Peel, J. D. Y. 1995. For Who Hath Despised the Day of Small Things? Missionary Narratives and Historical Anthropology. *Comparative Studies in Society and History,* 37: 581–607.

Peet, Richard, and Michael Watts, ed. 1996. *Liberation Ecologies: Environment, Development, and Social Movements.* London: Routledge.

Peluso, Nancy. 1992. *Rich Forests, Poor People: Resource Control and Resistance in Java.* Berkeley: University of California Press.

———. 1993. Coercing Conservation? The Politics of State Resource Control. *Global Environmental Change* (June): 199–218.

Pigg, Stacy. 1992. Inventing Social Categories through Place: Social Representations and Development in Nepal. *Comparative Studies in Society and History* 34, no. 3: 491–513.

Pocock, David. 1972. *Kanbi and Patidar.* Oxford: Clarendon Press.

Polanyi, K. 1944. *The Great Transformation: The Political and Economic Origins of Our Time.* Reprint, 1957. Boston: Beacon Press.

Postone, Moishe. 1996. *Time, Labor, and Social Domination.* Chicago: University of Chicago Press.

Prasad, Archana. 1994. Forests and Subsistence in Colonial India: A Study of Central Provinces, 1830–1945. Ph.D. diss., Jawaharlal Nehru University, New Delhi.

———. 1998. The Baigas: Survival Strategies and Local Economies in Colonial Central Provinces. *Studies in History* 14, no. 2: 325–48.

Prasad, R. R. 1994. *Pastoral Nomadism in Arid Zones of India: Socio-demographic and Ecological Aspects.* New Delhi: Discovery Publishing House.

PT (Parasnis Transcripts). Unpublished manuscripts in the Pune Archives.

Puri, G. S. 1949. The Problem of Land Erosion and Landslips in the Hoshiarpur Siwaliks. *Indian Forester* 75: 45–51.

Rabitoy, Neil. 1975. System v. Expediency: The Reality of Land Revenue Administration in the Bombay Presidency, 1812–20. *Modern Asian Studies* 9.

Rahnema, Majid. 1993. Participation. In *The Development Dictionary: A Guide to Knowledge as Power,* ed. W. Sachs, 116–31. London: Zed Books.

Rajan, Ravi S. 1994. Imperial Environmentalism: The Agendas and Ideologies of Natural Resource Management in British Colonial Forestry, 1800–1950. Ph.D. diss., University of Oxford, Oxford.

Rajputana Gazetteers. 1908. Compiled by K. D. Erskine. Reprint, Gurgaor (India): Vintage Books, 1922.

Rajvade, V. K., ed. 1909. *Marathyanchya itihasanchi Sadhanen—Khand Dahava.* Pune: Maharashtra Sahitya Parishad.

Rangan, H. 1993a. Of Myths and Movements: Forestry and Regional Development in the Garhwal Himalayas. Ph.D. diss., University of California, Los Angeles.

———. 1993b. Romancing the Environment: Popular Environmental Action in the Garhwal Himalayas. In *In Defense of Livelihood,* ed. Haripriya Rangan and John Friedman, 151–81. West Hartford, Conn.: Kumarian Press.

———.1995. Contested Boundaries: State Policies, Forest Classifications, and Deforestation in the Garhwal Himalayas. *Antipode* 27, no. 4: 343–62.

———. 1997. Property vs. Control: The State and Forest Management in the Indian Himalaya. *Development and Change* 28, no. 1: 71–94.

Rangarajan, Mahesh. 1994. Imperial Agendas and India's Forests: The Early History of Indian Forestry. *Indian Economic and Social History Review* 31, no. 2: 147–67.

———. 1996. *Fencing the Forest.* New Delhi: Oxford University Press.

———. 1998. The Raj and the Natural World: The War against Dangerous Beasts in Colonial India. *Studies in History,* n.s., 14, no. 2: 265–99.

Rangnekar, Sangeeta. 1994. Women Pastoralists, Indigenous Knowledge, and Livestock Production in Northern Gujarat. Pastoral Development Network Paper 36a. London: Overseas Development Institute.

Rao, A. L. 1988. Involvement of Weaker Section in Wasteland Afforestation Program. In *Wasteland Development for Fuelwood and Fodder Production,* 125–30. Dehra Dun: Forest Research Institute and Colleges, Government of India.

Rao, A. S. 1992. Climate, Climatic Changes, and Paleo-climatic Aspects of Rajasthan. In *Geographical Facets of Rajasthan,* ed. H. S. Sharma and M. L. Sharma, 38–44. Chandragar: Kuldeep Publications.

Rawat, A. S. 1983. *Garhwal Himalayas: A Historical Survey of the Political and Ad-ministrative History of Garhwal, 1815–1947.* Delhi: Eastern Book Linkers.

———. 1987. *Commentary on G. B. Pant's "Forest Problems in Kumaon."* Nainital: Gyanodaya Prakashan.

———. 1989. *History of Garhwal, 1358–1947: An Erstwhile Kingdom in the Himalayas.* New Delhi: Indus Publishing.

———, ed. 1991. *History of Forestry in India.* New Delhi: Indus Publishing.

Redclift, Michael, and Ted Benton, eds. 1994. *Social Theory and the Global Environment.* London: Routledge.

Ribbentrop, B. 1900. *Forestry in British India.* Reprint, 1989. New Delhi: Indus Publishing.

Ribot, Jesse C. 1997. Theorizing Access: Forest Profits along Senegal's Charcoal Commodity Chain. *Development and Change,* 29 (2): 307–41.

Richards, John F. 1987. Environmental Changes in Dehra Dun Valley, India: 1880–1980. *Mountain Research and Development* 7, no. 3: 299–304.

Richards, John F., and Elizabeth Flint. 1990. Long-Term Transformations in Sunderbans Wetlands Forests of Bengal. *Agriculture and Human Values* 7, no. 2: 17–33.

———. N.d. *Historic Land Use and Carbon Estimates for South and Southeast Asia, 1880–1980.* Ed. R. C. Daniels. Carbon Dioxide Information Analysis Center, Oak Ridge National Laboratory, Experimental Sciences Division, Publication no. 4174. Data from this study are available on the Internet.

Richards, John F., Edward S. Haynes, and James R. Hagen. 1985. Changing Land Use in Bihar, Punjab, and Haryana, 1850–1970. *Modern Asian Studies* 19, no. 4: 699–732.

Richards, John, and James Hagen. 1987. A

Century of Rural Expansion in Assam, 1870–1970. *Itinerario* 9: 193–209.

Richards, John, and Michelle McAlpin. 1983. Cotton Cultivating and Land Clearing in the Bombay Deccan and Karnatak, 1818–1920. In *Global Deforestation and the Nineteenth-Century World Economy,* ed. Richard Tucker and John Richards. Durham, N.C.: Duke University Press.

Richards, Paul. 1993. Cultivation: Knowledge or Performance? In *An Anthropological Critique of Development: The Growth of Ignorance.* London: Routledge.

Riwaj-i-Abpashi (Irrigation Customs). 1918. Compiled as part of the 1918 settlement of Kangra.

Robbins, Paul. 1994. Goats and Grasses in Western Rajasthan. Pastoral Development Network Paper 36a. London: Overseas Development Institute.

———. 1998a. Nomadization in Rajasthan, India: Migration, Institutions, and Economy. *Human Ecology* 26, no. 1: 87–112.

———. 1998b. Shrines and Butchers: Animals as Deities, Capital, and Meat in Contemporary North India. In *Animal Geographies,* ed. J. Wolch and J. Emel. London: Verso Press.

———. 1998c. Authority and Environment: Institutional Landscapes in Rajasthan, India. *The Annals of the Association of American Geographers* 88, no. 3: 410–35.

———. 1998d. Paper Forests: Imagining and Deploying Exogenous Ecologies in Arid India. *Geoforum* 29, no. 1: 69–89.

Rocheleau, Diane. 1985. Women, Trees, and Tenure: Implications for Agroforestry Research and Development. In *Land, Trees, and Tenure,* ed. J. B. Raintree, 79–120. Nairobi: ICRAF.

Rocheleau, Diane, B. Thomas-Slayter, and E. Wangari. 1996. Gender and Environment: A Feminist Political Ecology Perspective. In *Feminist Political Ecology: Global Issues and Local Experiences,* 3–26. London: Routledge.

Rolston, Holmes, III. 1988. *Environmental Ethics: Duties to and Values in the Natural World.* Philadelphia: Temple University Press.

Rose, Carol M. 1994. *Property and Persuasion: Essays on the History, Theory, and Rhetoric of Ownership.* Boulder: Westview Press.

Roseberry, W. 1989. *Anthropologies and Histories.* New Brunswick: Rutgers University Press.

Rutten, Mario. 1995. *Farms and Factories.* Delhi: Oxford University Press.

Saberwal, Vasant K. 1997. Bureaucratic Agendas and Conservation Policy in Himachal Pradesh, 1865–1994. *Indian Economic and Social History Review* 34: 465–98.

———. 1999. Pastoral Politics: Bureaucratics, Shepherds, and Conservation in the Western Himalaya. New Delhi: Oxford University Press.

Sachs, Wolfgang, ed. 1992. *The Development Dictionary: A Guide to Knowledge as Power.* London: Zed Books.

Sagar, Vidya, and K. Ahuja. 1993. *Economics of Goat Keeping in Rajasthan.* Jaipur: Institute of Development Studies.

Sahlins, Marshall. 1976. *Culture and Practical Reason.* Chicago: University of Chicago Press.

Said, Edward. 1979. *Orientalism.* New York: Vintage Books.

Saklani, A. 1986. *The History of a Himalayan Princely State: Change, Conflicts, and Awakenings: An Interpretive History of the Princely State of Tehri Garhwal, 1811–1949.* Delhi: Durga Publications.

Salzman, Phillip C. 1980. *When Nomads Settle: Processes of Sedentarization as Adaptation and Response.* New York: Praeger.

———. 1986. Shrinking Pasture for Rajasthani Pastoralists. *Nomadic Peoples* 20: 49–61.

Sanwal, Ram Dutt. 1976. *Social Stratification in Rural Kumaon.* Delhi: Oxford University Press.

Sarin, Madhu. 1993. From Conflict to Collaboration: Local Institutions in Joint Forest Management. JFM. Working Paper no. 14. New Delhi: SPWD/Ford Foundation.

———. 1995a. Joint Forest Management in India: Achievements and Unaddressed Challenges. *Unasylva: An International Journal of Forestry and Forest Industries* 46: 30–36.

———. 1995b. Regenerating India's Forests: Reconciling Gender Equity with Joint Forest Management. *Institute of Development Studies Bulletin* 26, no. 1: 83–91.

Sarin, Madhu, and C. Sharma. 1993. Experiments in the Field: The Case of PEDO in Rajasthan. In *Women and Wasteland Development in India,* ed. A. M. Singh and N. Burra. New Delhi: Sage Publications.

Saxena, N. C. 1994. Panchayats and Common Land Afforestation in India. In *Decentralization: Panchayats in the Nineties,* ed. A. Mukherjee. New Delhi: Vikas Publishing House.

Schama, Simon. 1995. *Landscape and Memory.* New York: Alfred Knopf.

Schlager, Edella, and Elinor Ostrom. 1992. Property Rights Regimes and Natural Resources: A Conceptual Analysis. *Land Economics* 68, no. 3: 249–62.

Schmink, Mariane, and Charles Wood. 1992. *Contested Frontiers in Amazonia.* New York: Columbia University Press.

Schwartzberg, Joseph E. 1992. *A Historical Atlas of South Asia.* Delhi.

Scott, James C. 1976. *The Moral Economy of the Peasant: Rebellion and Subsistence in South-east Asia.* New Haven: Yale University Press.

———. 1985. *Weapons of the Weak: Everyday Forms of Resistance.* New Haven: Yale University Press.

———. 1998. *Seeing Like a State.* New Haven: Yale University Press.

Sears, Paul. 1935. Deserts on the March. Norman: University of Oklahoma Press.

Sen, Amartya. 1966. Peasants and Dualism with or without Surplus Labour. *Journal of Political Economy* 74: 425–50.

———. 1981. *Poverty and Famines: An Essay on Entitlement and Deprivation.* Oxford: Oxford University Press.

Sen, Asok. 1987. Subaltern Studies: Capital, Class and Community. In *Subaltern Studies V: Writings in South Asian History and Society,* ed. Ranajit Guha, 203–35. Delhi: Oxford University Press.

Sen, G., ed. 1992. *Indigenous Vision: Peoples of India Attitudes to the Environment.* New Delhi: Sage.

Shah, Anup. 1998. *Ecology and the Crisis of Overpopulation: Future Prospects for Global Sustainability.* Cheltenham, U.K.: Edward Elgar.

Shah, Vimal, and C. H. Shah. 1974. *Resurvey of Matar Taluka.* Bombay: Vora Publishers.

Shah, Vimal, C. H. Shah, and Sudershan Iyengar. 1990. *Agricultural Growth with Equity.* Delhi: Concept.

Sharma, J. P. 1992. *Peasant-Base of Indian Democracy.* Jaipur: RBSA Publishers.

Sheppard, Eric, and Trevor Barnes. 1990. *The Capitalist Space Economy.* London: Unwin Hyman.

Shiva, Vandana. 1989a. *Staying Alive: Women, Ecology, and Development.* London: Zed Books.

———. 1989b. *The Violence of the Green Revolution: Ecological Degradation and Political*

Conflict in Punjab. Dehra Dun: Research Foundation for Science and Ecology.

———. 1991. *Violence of the Green Revolution: Third World Agriculture, Ecology, and Politics*. London: Zed Books.

Shiva, Vandana, and J. Bandyopadhyay. 1986a. Environmental Conflicts and Public Interest Science. *Economic and Political Weekly* 21, no. 2: 84–90.

———. 1986b. *Chipko: India's Civilizational Response to the Forest Crisis*. New Delhi: Intach.

Shiva, Vandana, and M. Mies. 1993. *Ecofeminism*. London: Zed Books.

Singh, Chatrapati. 1986. *Common Property and Common Poverty: India's Forests, Forest Dwellers, and the Law*. Delhi: Oxford University Press.

Singh, Chetan. 1991. Humans and Forests: The Himalaya and the *Terai* during the Medieval Period. In *History of Forestry in India*, ed. A. S. Rawat, 163–78. New Delhi: Indus Publishing.

———. 1995. Forests, Pastoralists, and Agrarian Society in Mughal India. In *Nature, Culture, Imperialism: Essays on the Environmental History of South Asia,* ed. David Arnold and Ramachandra Guha. Delhi: Oxford University Press.

———. 1998. *Natural Premises: Ecology and Peasant Life in the Western Himalaya, 1800–1950*. Delhi: Oxford University Press.

Singh, Munshi Hardyal. 1894. *The Castes of Marwar*. Jodhpur: Books Treasure.

Sinha, Subir, Shubhra Gururani, and Brian Greenberg. 1997. The "New Traditionalist" Discourse of Indian Environmentalism. *Journal of Peasant Studies* 24, no. 3: 65–99.

Sivaramakrishnan, K. 1995. Imagining the Past in Present Politics: Colonialism and Forestry in India. *Comparative Studies in Society and History* 37, no. 1: 3–40.

———. 1996. Forests, Politics, and Governance in Bengal, 1794–1994. Vols. 1–2. Ph.D. diss., Yale University.

———. 1997. A Limited Forest Conservancy in Southwest Bengal, 1864–1912. *Journal of Asian Studies* 50, no. 1: 75–112.

———. 1998. Modern Forestry: Trees and Development Spaces in West Bengal. In *The Social Life of Trees: Anthropological Perspectives on Tree Symbolism,* ed. Lara Rival. Oxford: Berg.

Skaria, Ajay. 1997. Shades of Wildness: Caste, Tribe, and Gender in Western India. *Journal of Asian Studies* 56, no. 3 (August).

———. 1999. *Hybrid Histories: Forests, Frontiers, and Wildness in Western India*. New Delhi: Oxford University Press.

Smith, Neil. 1998. Nature at the Millennium: Production and Re-enchantment. In *Remaking Reality: Nature at the Millennium,* ed. B. Braun and N. Castree, 271–85. London: Routledge.

Smythies, E. A. 1939. Erosion and Floods: Problems of Soil and Water Conservation in the United Provinces. *Indian Forester* 75: 354–60.

Soja, Edward. 1989. *Postmodern Geographies: The Reassertion of Space in Critical Social Theory*. London: Verso.

Somanathan, E. 1991. Deforestation, Property Rights, and Incentives in Central Himalaya. *Economic and Political Weekly,* 26 January, 37–46.

Soper, K. 1995. *What Is Nature? Culture, Politics, and the Non-human*. Oxford: Blackwell.

Sopher, David E. 1975. Indian Pastoral Castes and Livestock Ecologies: A Geographic Analysis. In *Pastoralists and Nomads in South Asia,* ed. L. S. Leshnik and G.-D. Sontheimer. Wiesbaden: Otto Harrassowitz.

Sorenson, John. 1993. *Imagining Ethiopia:*

Struggles for History and Identity in the Horn of Africa. New Brunswick: Rutgers University Press.

SPD. 1931–1935. Govind S. Sardesai, ed. *Selections from the Peshwa Daftar.* 46 vols. Bombay: Government Central Press.

Sponsel, Leslie E., Thomas N. Headland, and Robert C. Bailey, eds. 1996. *Tropical Deforestation: The Human Dimension.* New York: Columbia University Press.

Srinivas, M. N. 1994. *The Dominant Caste and Other Essays.* Oxford: Oxford University Press.

Srivastava, V. K. 1991. Who Are the Raikas/Rabaris? *Man in India* 71, no. 1: 279–304.

SSRPD. 1902–1911. G. C. Vad, comp. *Selections from the Satara Raja's and Peshwa Diaries.* 9 parts. Pune: Deccan Vernacular Translation Society.

Stebbing, E. P. 1922. *The Forests of India.* Vol. 1. London: Bodley Head.

———. 1932. *The Forests of India.* Vol. 3. London: Bodley Head.

Stein, Burton. 1989. Eighteenth Century India: Another View. *Studies in History,* n.s., 5, no. 1: 1–26.

Stokes, E. 1959. *English Utilitarians and India.* Oxford: Oxford University Press.

———. 1978. *The Peasant and the Raj: Studies in Agrarian Society and Peasant Rebellion in Colonial India.* Cambridge: Cambridge University Press.

Sykes, William H. 1835. On the Land Tenures of the Deccan. *Journal of the Royal Asiatic Society of Great Britain and Ireland* 2: 206–33.

Taussig, Michael. 1980. *The Devil and Commodity Fetishism in South America.* Chapel Hill: University of North Carolina Press.

Taylor, Charles. 1993. To Follow a Rule. In *Bourdieu: Critical Perspectives,* ed. Craig Calhoun, Edward LiPuma, and Moishe

Postone. Chicago: University of Chicago Press.

Thapar, Romila. 1978. *Ancient Indian Social History, Some Interpretations.* New Delhi.

Thompson, E. P. 1971. The Moral Economy of the English Crowd in the Eighteenth Century. *Past and Present* 50: 73–136.

Thompson, John B. 1984. *Studies in the Theory of Ideology.* Oxford: Basil Blackwell.

Traill, G. W. 1828. A Statistical Sketch of Kumaon. *Asiatic Researches.* Vol. 16. Reprint, 1980. New Delhi: Indus Publishing.

Troup, R. S. 1922. *Silviculture of Indian Trees.* Vols. 1–3. Reprint, 1976. Dehra Dun: Indian Forest Research Institute.

Tucker, Compton J., Harold Dregne, and Wilbur Newcomb. 1991. Expansion and Contraction of the Sahara Desert from 1980 to 1990. *Science* 253: 299–301.

Tucker, Richard P. 1982. The Forests of the Western Himalayas: The Legacy of British Colonial Administration. *Journal of Forest History,* 26 (3): 112–23.

———. 1983. The British Colonial System and the Forests of the Western Himalayas, 1815–1914. In *Global Deforestation and the Nineteenth-Century World Economy,* ed. R. P. Tucker and J. F. Richards, 146–66. Durham, N.C.: Duke University Press.

———. 1985. The Evolution of Transhumant Grazing in the Punjab Himalaya. *Mountain Research and Development* 6: 17–28.

———. 1988. The British Empire and India's Forest Resources: The Timberlands of Assam and Kumaon, 1914–1950. In *World Deforestation in the Twentieth Century,* ed. J. F. Richards and R. P. Tucker, 91–111. Durham, N.C.: Duke University Press.

———. 1989. The Depletion of India's Forests under British Imperialism: Planters, Forests, and Peasants in Assam and Kerala. In *The Ends of the Earth: Perspec-*

tives on *Modern Environmental History*, ed. D. Worster. Cambridge: Cambridge University Press.

———. 1991. The Evolution of Transhumant Grazing in the Punjab Himalaya. In *History of Forestry in India*, ed. A. S. Rawat, 215–40. New Delhi: Indus Publishing.

Turner, S. 1800. *An Account of an Embassy to the Court of Teshoo Lama in Tibet*. London: Bulmer and Row.

United Provinces of Agra and Oudh Forest Manual. 1936. Sixth Edition. Allahabad: Government Press.

Vandergeest, Peter. 1996. Property Rights in Protected Areas: Obstacles to Community Involvement as a Solution in Thailand. *Environmental Conservation* 23, no. 4.

———. 1997. Rethinking Property. *The Common Property Resource Digest* 41: 4–6.

Vandergeest, Peter, and Nancy Peluso. 1995. Territorialization and State Power in Thailand. *Theory and Society* 24: 385–426.

Voloshinov, V. N. 1973. *Marxism and the Philosophy of Language*. Trans. L. Matejka and I. R. Titunik. Cambridge: Harvard University Press.

Wade, Robert. 1988. *Village Republics: Economic Conditions for Collective Action*. Cambridge: Cambridge University Press.

Walton, H. G. 1910. *British Garhwal: A Gazetteer Being Volume XXXVI of the District Gazetteers of the United Provinces of Agra and Oudh*. Reprint, 1989. Dehra Dun: Natraj Publishers.

Warren, D. Michael, Jan Slikkerveer, and David Brokensha, eds. 1995. *The Cultural Dimension of Development: Indigenous Knowledge Systems*. London: Intermediate Technology Publications.

Washbrook, David. 1978. Economic Development and Social Stratification in Rural Madras: The "Dry Region," 1878–1929.

In *The Imperial Impact: Studies in the Economic History of Africa and India,* ed. Clive Dewey and A. G. Hopkins. London: Athlone Press.

———. 1988. Progress and Problems: South Asian Economic and Social History, c. 1720–1860. *Modern Asian Studies* 22, no. 1: 57–97.

Weber, J. 1988. *Hugging the Trees: The Story of the Chipko Movement*. New Delhi: Viking.

Westphal-Hellbusch, S. 1975. Changes in the Meaning of Ethnic Names as Exemplified by the Jat, Rabari, Bharvad, and Charan in Northwestern India. In *Pastoralists and Nomads in South Asia,* ed. L. S. Leshnik and G.-D. Sontheimer. Wiesbaden: Otto Harrassowitz.

Whitcombe, E. 1972. *Agrarian Conditions in Northern India: The United Provinces under British Rule, 1860–1900*. Vol. 1. Berkeley and Los Angeles: University of California Press.

Whitehead, A. 1984. Men and Women, Kinship and Property: Some General Issues. In *Women and Property: Women as Property,* ed. R. Hirschon. London: Croom Helm.

Williams, G. R. C. 1874. *Historical and Statistical Memoir of Dehra Doon*. Reprint, 1985. Dehra Dun: Natraj Publishers.

Williams, Raymond. 1976. *Keywords: A Vocabulary of Culture and Society*. London: Fontana.

Wilson, Ken. 1995. A Water Used to Be Scattered in the Landscape: Local Understandings of Soil Erosion and Land Use Planning in Southern Zimbabwe. *Environment and History* 1, no. 3: 281–96.

Wiser, William, and C. Wiser. 1963. *Behind Mud Walls: 1930–1960*. Berkeley: University of California Press.

Woodman, D. 1969. *Himalayan Frontiers: A Political Review of British, Chinese, Indian, and Russian Rivalries.* London: Barrie and Rockliff.

Worster, David. 1979. *Dustbowl: The Southern Plains in the 1930s.* New York: Oxford University Press.

Wrigley, C. C. 1978. Neo-mercantile Policies and the New Imperialism. In *The Imperial Impact: Studies in the Economic History of Africa and India,* ed. C. Dewey and A. G. Hopkins. University of London, Institute of Commonwealth Studies, Commonwealth Papers 21. London: Athlone Press.

Wrigley, Christopher. 1996. *Kingship and State: The Buganda Dynasty.* Cambridge: Cambridge University Press.

Xenos, Nicholas. 1989. *Scarcity and Modernity.* New York: Routledge.

Young, Iris. 1990. The Ideal of Community and the Politics of Difference. In *Feminism/Postmodernism,* ed. L. J. Nicholson. New York: Routledge.

Zaidi, A. Moin, ed. 1985. *A Tryst with Destiny: A Study of Economic Policy Resolutions of the Indian National Congress Passed during the Last 100 Years.* New Delhi.

Contributors

Arun Agrawal is Assistant Professor of Political Science at Yale University. His research examines pastoralists and forest-dependent communities in South Asia. He has published several articles on migrant shepherds in Rajasthan, forest panchayats in Uttar Pradesh, and indigenous knowledge. His first book, *Greener Pastures: Politics, Markets, and Community among a Migrant Pastoral People,* has been published by Duke and Oxford University Presses. He has also published a short monograph with ICS Press, *Decentralization in Nepal.*

J. Mark Baker was Assistant Professor of Environmental Policy and Natural Resource Management in the Environmental Studies Program at the University of North Carolina, Asheville. His research interests concern community-based resource management. He has studied community forestry and local irrigation management in Bihar and Himachal Pradesh, India, and the evolution of watershed institutions in the state of California. He is currently affiliated with the nonprofit organization Forest Community Research and is conducting applied research on community-based forestry in the United States.

Molly Chattopadhyay is a sociologist in the Agricultural Science Unit of the Indian Statistical Institute, Calcutta, with research interests in women and work, in both formal sector employment and agriculture, and in rural livelihoods and well-being in West Bengal and Bihar.

Vinay Gidwani was, at the time of writing, an Izaak Killam Postdoctoral Fellow at the University of British Columbia, Vancouver, B.C., Canada. Since September 1999, he has been at the University of Minnesota, Twin Cities, as Assistant Professor in the Department of Geography and the Institute for Global Studies. His areas of academic interest are development economics, agrarian studies, and political ecology.

Sumit Guha is Professor of History at Brown University. His latest book, *Environment and Ethnicity in India, 1200–1991,* was published in 1999 by

Cambridge University Press. He has also published books with Oxford University Press.

Shubhra Gururani is Assistant Professor of Anthropology at York University, Canada, where she teaches political ecology, anthropology of development, and feminist ethnography. She is currently working on a manuscript that ethnographically explores environmental and development politics in the central Himalaya. Her recent project is on environmental and territorial politics in Third World cities.

Cecile Jackson is a Senior Lecturer in the School of Development Studies, University of East Anglia. She has researched gender and agrarian change since the mid-1970s in Nigeria, Zimbabwe, and later in Bihar and West Bengal, focusing on environmental issues in particular but also on reproductive decision making and rural development interventions.

David Ludden is Associate Professor of History and South Asian Regional Studies, University of Pennsylvania. He has written *Peasant History in South India* (Princeton 1985, Oxford 1989) and *An Agrarian History of South Asia* (Cambridge University Press, 1999).

Haripriya Rangan is Lecturer in the Department of Geography and Environmental Science, Monash University. She has published several articles on the politics and history of forest conservation and regional development in the Uttarakhand region of India. She is currently working on a research project on petty commodity extraction from rural state lands in South Africa.

Paul Robbins is Assistant Professor of Geography at Ohio State University. He has published widely on the economy and ecology of pastoralism, the impact of competing knowledges on resource policy, and the role of normative social institutions in creating the patchwork landscapes of semi-arid India. His most recent research explores human and wild animal adaptations to the radically altered landscapes created by invasive plant species in the Indian Aravalli.

Vasant K. Saberwal is a researcher at the Center for Research and Action on Biodiversity, Pune. He has published in the areas of wildlife conservation and grazing policy. His book *Pastoral Politics: Shepherds, Bureaucrats, and Conservation in the Western Himalaya, 1865–1994* has been published by Oxford University Press.

James C. Scott is the Eugene Meyer Professor of Political Science, Professor of Anthropology, and Director of the Program in Agrarian Studies, Yale University. He is the author of *Moral Economy of the Peasant* (1976), *Weapons of the Weak: Everyday Forms of Peasant Resistance* (1985), and *Domination and the Arts of Resistance: Hidden Transcripts* (1990). His most recent book, *Seeing Like a State,* has been published by Yale University Press (1998).

K. Sivaramakrishnan is Assistant Professor in the Department of Anthropology at the University of Washington, Seattle. He has published several articles on the colonial and contemporary politics of forest management in Bengal, eastern India. His book *Modern Forests: Statemaking and Environmental Change in Colonial Eastern India* was jointly published by Stanford University Press and Oxford University Press in 1999.

Ajay Skaria is Assistant Professor of History at the University of Minnesota, Twin Cities. He is a member of the Subaltern Studies Collective and has published several articles based on his research on western India. His book *Hybrid Histories: Forests, Frontiers, and Oral Traditions in Western India* was published in 1999 by Oxford University Press.

Jenny Springer is currently working with the World Wildlife Fund, U.S.A. She is also a Ph.D. candidate in Sociocultural Anthropology at the University of Chicago. Her research interests include agrarian change in the Philippines and India, discourses of development, and community-based resource management. She has recently finished writing *Forestry for Sustainable Rural Development: A Review of Ford Foundation Supported Community Forestry Programs in Asia,* to be published by the Ford Foundation.

Darren C. Zook was Assistant Professor in Asian History at Claremont College. He works on the social and intellectual history of agricultural development in southern India and is starting a long-term comparative project on the meanings of hunger.

Index

210, 214, 251–263; agents of, 89, 91, 93, 99, 104; community, 115, 191–193, 196, 213; critiques of, 86, 195, 252, 274; as different from economic growth, 254; discourses of, 86, 96, 194, 199, 253; global, 251; goals, 95; grassroots, 253; ideologies of, 89–91; Indian paradigm for, 192, 276; "maldevelopment," 24; mission, 103; participatory, 191, 194; "people-centered," 191; planning, 197, 202, 213; programs, 87, 150, 169n 4; progressive, 192; regimes, 251, 268–276; rural, 115; sustainable, 191; targets for, 89, 99, 104, 195; "underdevelopment," 86

Dikku (Hindu incomers), 147, 158, 160, 168n 1

Discipline, 195, 197

Divorce, 160, 162, 165

Doab, 37–38

Dominant policy phases, 26

Dowry, 54, 156, 162

Drought, 37, 77–79, 198, 220, 262

Drought Prone Areas Program, 209

Duke University project, 13

East Africa, 236

Ecofeminism, 8, 14, 147–148; discourses of, 149

Ecologicalism, 6

Ecology: human, 218; literature, 78; science of, 2, 81

Economic: change, 222, 224; growth, 194; liberalism, 114, 127; sociology, 217; stratification, 217

Ecopopulists, 153

"Ecosystem people," 133, 141

Education, 184, 187; lack of, 185–187

Elites, 65, 203, 205, 207–209, 220, 253; and coalitions, 210; local 18, 42, 44, 200; scholarship, 4

Endogamy, 124

Environment, 2, 107; histories of, 9; images of, 92; moral, 109, 128; ontological status of, 2

Environmental: conflict, 13, 14, 20; history, 2, 5, 10–16, 132, 134, 265, 275–276; man-

agement, 8; policy, 6, 19–20, 149, 189; politics, 2, 14, 19; scarcity, 158; studies, 2, 3, 5, 10–11, 19, 127

Environmental catastrophism, 14, 78

Environmental change, 7, 13, 23–24, 45, 196, 259–260, 262–263; gendered views of, 155; local perceptions of, 152, 153, 155

Environmental degradation, 7, 12, 23, 74, 76, 107, 152–153, 158, 170, 265; discourses about, 68–70, 77, 78, 84n 20; gendered nature of, 7; historical patterns of, 114; ideologies of, 173; political opposition to, 87

Environmentalism, 6, 77, 155, 252–255, 263, 273–274; activists, 2, 21n 12, 151; in developed versus developing countries, 12, 14–15; in India, 10, 190n 8, 258; populist, 147; radical, 147, 149; and social justice, 15, 148; Western, 45, 147

Erosion, 42, 68–78, 152, 153, 165; ambivalent views on, 153

Ethnicity, 147, 160–161, 168n 1; discourses of, 167. *See also* Identity: ethnic

"Ethnographic thickness," 149

European expansion, 14

Expertise: as gendered, 166

Exports, 34, 37, 41, 43

Fallows, 37, 198, 206, 208, 210; shortening of, 241

"False consciousness," 112

Family planning, 126

Famine, 37, 43, 107, 109, 112, 115, 118, 120, 126, 236, 262; in Africa, 111, 114; Bengal famine of 1942–1943, 110–111, 116–117; Bengal famine of 1770, 129n 9; Codes, 117, 130n 20; "Great Famine" of 1876–1878, 115–116; Madras famine of 1876–1878, 120; narratives of, 107–110, 113–117, 121, 125, 128, 267; prevention of, 59, 113–114, 121, 251, 266; responsibility for, 112–120; scholarship on, 13, 17, 127; and social justice, 127–128

Human rights, 87, 252

Hunger, 114, 262; chronic, 122; representations of, 13, 17. *See also* Famine

Hypergamy, 124, 162. *See also* Marriage

Ideal-typical contructs, 5, 7

Identity: caste, 157, 162; cultural, 150; discourses, 147, 160; ethnic, 149, 154–155, 160, 165–167; gender, 155, 165–167; group, 7, 210; and livelihood, 149; multiple, 155; national, 254; politics of, 6; self-, 218; social, 8, 14, 155

IMF. *See* International Monetary Fund

Imperial Forest Service, 41, 62

Imperialism, 111

Imports, 34, 40; tariffs on, 43

Indian Constitution, 126

Indian Famine Relief Fund, 119, 129n 14

Indian Forest Act of 1878, 84

Indian mutiny, 32

Indian studies, 10

Indigenous, 7–9, 12, 19–20, 21n 18, 89, 181, 259; knowledge, 8, 87, 96, 103

Individualism, 212; and individualization, 25, 43

Indo-Gangetic Plains, 4, 16, 28–32, 37, 40, 82

Industrialization, 32; forest-based, 44

Infanticide, female, 162

Innovation, 105, 166

Insiders, 117, 134, 169n 7

Institution building, 214

Integrated Pest Management, 102

Intercropping, 165–166

International Monetary Fund (IMF), 241, 252

Irrigation, 4, 6, 37–38, 48, 51, 55, 57, 65, 86, 88, 92, 118, 165–166, 196, 216, 218, 220, 224, 230, 238, 241, 243, 259, 261–262; rights, 58, 139; unexpected effects of, 238

Irrigation Department, 92

Jats, 200, 202, 207, 209

Jharkand, 150–151, 154, 156, 160, 168n 4; Co-ordination Committee, 150; movement, 150–152, 160, 167; Mukti Morcha, 150

"Jungle people," 155–156

Kanbi, 218

Kangra Fort, 53

Katoch rule, 49, 51

Kin: groups, 51; hierarchies, 182; networks, 181; status, 184

Kingdoms, 28–31, 112

Kohl, 155

Kohli (water master), 66n 6

Kolam, 153

Kol rebellion, 151

Kuhl (irrigation system), 51, 57–59, 65. *See also* Irrigation

Kumaon, 28–35, 38–44, 48, 172–190

Labor, 162–163, 200, 203, 218, 237; corvée, 221; demand for, 38, 44, 59, 239–241; displacement, 239; diversification, 230, 237, 241; division of, 193; domestic, 161, 165, 169n 11; family, 217, 242; forced, 136–137, 143; as mediating culture and nature, 218; migrant, 163; piece rates, 238; "power," 242; reproductive, 161, 164; reserves, 218; shunning certain types of, 230, 237–238; supply conditions of, 31, 239; surpluses, 230; wage, 156–157, 161, 164–165, 169n 11, 179, 239, 241

Lalkhandis, 150

Laloo government, 152

Lambedar (village tax collector), 59

Lancashire lobby, 32, 41

Land: absentee ownership of, 55, 65; accumulation, 54, 161; arid, 194; assessment rates, 53; Barhi, 161, 163, 165; *Chos* 73–75; as collateral, 225; communal, 206, 208; demarcation, 176; *Garha,* 163–164; and gender, 157, 159, 160, 162, 164; *Gocher* (grazing land), 205, 208; grants, 49, 51; inherited, 161, 224; management, 17; as medium of exchange,

174, 198; claims on, 147; collective-choice rights, 174; competition for, 13, 138, 141, 142, 144, 157, 167, 172, 186, 189, 262–263; degradation of, 2; exclusion rights, 174; as focus of study, 2; gendered use of, 172, 183; human impact on, 2; internal-use patterns, 173; management of, 5, 7, 9, 15, 62, 65, 145, 153, 158, 173, 192; perceptions of, 148, 154; politicization of, 154–155; regulations regarding use of, 62, 70, 73, 82, 133–134, 138, 155, 176; stewards of, 219; as term unacceptable to environmentalists, 147

Nature, 2, 9, 11–12, 19, 265; as autonomous, 1; conquest of, 115; cultural entitlements to, 149; as feminine, 23, 148, 265, 267; harmony with, 167; historical understanding of, 275; and human nature, 218; as idea, 148–149, 160, 168; ideas of, 3, 19, 148, 167, 186, 232, 243, 265–276; as machine, 275; sacralization of, 153; social construction of, 3, 266; state of, 20n 4; subordination of, 274; unpredictability of, 230, 232, 241, 243; work of, 219, 230, 232, 243

Nehru, Jawaharlal, 126, 251, 258, 272

Nepal, 29

Ninth Plan paper, 133

Nongovernmental organizations (NGOs), 146, 213, 252–253

Northwest Frontier Provinces, 65

Occupational markers, 15

Opium, 34

The Opium War, 39

Oral history, 110, 113–114, 128n 5

Oran (village forest-pasture), 205, 210, 212

Orientalism, 23, 270

Outsiders, 8, 20, 108, 117, 130n 21, 134, 152–153, 169n 7, 187

Overgrazing, 5, 17, 30, 68–69, 75, 77, 82

"Overpopulation," 124–126

Pakhtun, 256

Panchayat (village forest), 171–172, 182–183,

187; committees, 182, 183. *See also* Village, councils

Pashm (cashmere wool), 33, 40

Pastoralists, 1, 5–6, 18, 69, 75, 82, 136, 192, 212, 221; adaptive strategies of, 6, 200, 206; as cause of degradation of environment, 69; diversity in practices of, 197, 199–214; images of, 191, 196, 198, 202; as male, 204; mobile, 1, 18, 196, 200; nontraditional, 202; settlement of, 197, 200–201

Patidars, 220

Patriarchy, 8, 150–151, 160, 181–183; patriarchal complex, 161–162; Western, 23

Patrilineal inheritance, 161

Patron-client relationships, 166, 209, 221

Patta (deed), 50, 51, 245n 13

Pax Brittanica, 269

Phulchi, 149–168

Polarization theory, 217–218, 242–243

Political: alliances, 213; ecology, 21n 12, 205, 209; economy, 191, 195–196, 253, 212; independence, 152, 205; legitimacy, 49, 51; repression, 150–152

Political-ecological research, 7

Pollarding, 152

Pollution, 259

Polygamy, 124

Popular culture, 125, 217

Popular media, 83n 1, 109–110, 119–120, 125, 152, 200

Population: density, 134; displacement, 262; growth, 3, 74, 136, 154, 260

Postmodern historiography, 108

Poverty, 114, 117, 125–126, 258, 273; causes of, 123–124, 128; chronic, 115, 119; as cultural ideal, 124; and environmental impacts, 154; eradication programs, 126; mass, 120; representations of, 17, 116, 124, 126, 266; poverty-ridden nations, 108, 114, 125

Power, 92, 133, 183, 188, 210, 234, 274; balance of, 150; constellations of, 149; expert, 193; layers of, 181; local, 193; shifts, 252; struggles over, 172

"Woman," 7–10, 18, 19–20, 148

Women: agency of, 162; control over, 151, 161–162; and knowledge, 161, 163, 166, 182, 186–187; and labor, 162–166, 169n 9, 177, 179, 181–182, 184, 204, 211, 237; and land, 157–160, 162, 164; as marginal, 159, 196; movement, 252; portrayals of, 45n 1; position of, 150–151, 182–184, 204; resistance of, 162; seclusion of, 162–163; as traditional, 211

Work ethic, 94

World Bank, 88, 96, 99, 252

Yadavs, 152

Yagya, 154

Yamuna River, 37

Zamindar (landholder), 40

Library of Congress Cataloging-in-Publication Data
Agrarian environments : resources, representations,
and rule in India / edited by Arun Agrawal and K.
Sivaramakrishnan ; foreword by James C. Scott.
p.   cm.
Includes bibliographical references and index.
ISBN 0-8223-2555-1 (cloth : alk. paper)
ISBN 0-8223-2574-8 (paper : alk. paper)
1. Land tenure—India.   2. Economic development—
Environmental aspects—India.   3. Agriculture—Eco-
nomic aspects—India.   4. Agriculture and state—India.
I. Agrawal, Arun, 1962–   II. Sivaramakrishnan, K., 1957–
HD876 .A553   2000
333.76'0954—dc21
00-027442